UTILITY
RABBIT KEEPING

FOR FOOD AND FUR

WITH HINTS ON
EXHIBITION RABBIT KEEPING

BY

PERCIVAL BRETTON

Tenth Edition

British Library Cataloguing-in-Publication Data
A catalogue record for this book is available from
the British Library

Cuniculture (Rabbit Farming)

Cuniculture is the agricultural practice of breeding and raising domestic rabbits, usually for their meat, fur, or wool. Some people however, called *rabbit fanciers*, practice cuniculture predominantly for exhibition. This differs from the simpler practice of keeping a single or small group of rabbits as companions, without selective breeding, reproduction, or the care of young animals. The distribution of rabbit farming varies across the globe, and while it is on the decline in some nations, in others it is expanding.

Domestication of the European rabbit rose slowly from a combination of game-keeping and animal husbandry. Among the numerous foodstuffs imported by sea to Rome during her domination of the Mediterranean were shipments of rabbits from Spain, they then spread across the Roman Empire. Rabbits were kept in both walled areas as well as more extensively in game-preserves. In the British Isles, these preserves were known as warrens or garths, and rabbits were known as coneys, to differentiate them from the similar hares (a separate species). The term warren was also used as a name for the location where hares, partridges and pheasants were kept, under the watch of a game keeper called a warrener. In order to confine and protect the rabbits, a wall or thick hedge might be constructed around the warren, or a warren might be established on an island.

Rabbits were typically kept as part of the household livestock by peasants and villagers throughout Europe. Husbandry of the rabbits, including collecting weeds and grasses for fodder, typically fell to the children of the household or farmstead. These rabbits were largely 'common' or 'meat' rabbits and not of a particular breed, although regional strains and types did arise. Some of these strains remain as regional breeds, such as the *Gothland* of Sweden, while others, such as the *Land Kaninchen*, a spotted rabbit of Germany, have become extinct. Contrary to intuitive sense, it was the development of refrigerated shipping vessels that led to the eventual collapse of European trading in rabbit meat. Such vessels allowed the Australians to harvest and more importantly, sell their over-population of feral rabbits.

With the rise of scientific animal breeding in the late 1700s, led by Robert Bakewell (among others), distinct livestock breeds were developed for specific purposes. Rabbits were among the last of the domestic animals to have these principles applied to them, but the rabbit's rapid reproductive cycle allowed for marked progress towards a breeding goal in a short period of time. Additionally, rabbits could be kept on a small area, with a single person caring for over 300 breeding does on an acre of land. This led to a short-lived eighteenth century 'boom' in rabbit breeding, selling, and speculation, when a quality breeding animal could bring $75 to $200. (For comparison, the average daily wage was approximately $1.00.) The final leg of deliberate rabbit breeding – beyond meat, wool & fur - was the breeding of 'fancy' animals as pets and curiosity. The term 'fancy' was originally applied to long eared 'lop' rabbits, as the lop rabbits were the first rabbits bred for exhibition. They were first admitted to agricultural shows in England in the 1820s, and in 1840 a club was formed for the promotion and regulation of exhibitions for 'Fancy Rabbits'.

In 1918, a new group formed for the promotion of fur breeds, originally including only Beverans and Havana breeds – now known as the 'British Rabbit Council.' In more recent years and in some countries, cuniculture has come under pressure from animal rights activists on several fronts. The use of animals, including rabbits, in scientific experiments has been subject to increased scrutiny in developed countries. Meanwhile, various rescue groups under the House Rabbit Society umbrella have taken an increasingly strident stance against any breeding of rabbits (even as food in developing countries) on the grounds that it contributes to the number of mistreated, unwanted or abandoned animals. Some of these organizations have promoted investigation and prosecution of rabbit raisers on humanitarian concerns.

Another important factor has been the growth of homesteaders and small holders, leading to the rise of visibility of rabbit raisers in geographic areas where they have not been present previously. This has led to zoning conflicts over the regulation of butchering and waste management. Ironically, many homesteaders have turned to cuniculture due to concerns over commercial rearing of other animals, namely cows, chickens and pigs, as well as a desire for more self-sufficient living. Conflicts have also arisen with House Rabbit Society organizations as well as ethical vegetarians and vegans concerning the use of rabbits as meat and fur animals rather than as pets. The specific future direction of cuniculture is unclear, but does not appear to be in danger of disappearing in any particular part of the world. The variety of applications, as well as the versatile utility of the species, appears sufficient to keep rabbit raising a going concern across the planet.

THE "SMALLHOLDER" LIBRARY
No. IX

UTILITY RABBIT KEEPING

CONTENTS

LIST OF PLATES

INTRODUCTION

SOME twelve years ago utility rabbit-keeping was, to all intents and purposes, unknown in this country. The youngsters kept rabbits as pets, of course, and crowds of adults kept rabbits for exhibition purposes, but none of us realised that Bunny had any utility value.

What a difference to-day! In thousands of homes throughout the country the useful qualities that we now know rabbits to possess are actually helping the house-wife to balance her weekly budget! In thousands of homes rabbits have proved themselves a most profitable side-line, rivalling in that respect chickens.

Yes, it is an actual fact that rabbit-keeping can to-day be almost as profitable a hobby as poultry-keeping. It actually scores over poultry-keeping in one way; it can be carried on in much more confined areas.

But, it may be argued, chickens have a double purpose; they not only lay eggs, but when they are too old to lay they shape well in roast or boiled form on the table. Well, nowadays—in the last year or two—rabbits have a double purpose, too. *They* shape well on the table in boiled or roast form, are, in fact, a regular item on every menu in town and country, and, in addition, their skins can be sold for anything from a few pence to seven shillings apiece.

That will, no doubt, surprise a good many of you who have not kept abreast of the times, who imagine that the only source for the disposal of rabbit skins is *via* the rag-a-bone man. Well, then, here's another surprise for you. Nine out of ten of the fur coats that women wear to-day are rabbit-skin coats, made from the skins of tame, hutch-reared rabbits that quite likely first saw the light of day, spent the whole of their lives, in some British backyard!

Obviously the demand for good-quality skins is there-

9

fore enormous ; obviously pelt production—they call rabbit skins " pelts " in the trade—has become a definite industry.

What is even more important, it is one of the few industries which has not become overcrowded. The demand for good-quality pelts is far and away in excess of the supply ; no rabbit-keeper need ever fear that he will not be able to sell his produce if only he will bear in mind these three important factors :

(1) He must keep a breed of rabbit that has the right sort of pelt—one of the fur-bearing breeds, in fact. There is practically no demand for common rabbit skins, even for the skins of those old favourites, the Belgian Hare and the Flemish Giant. The latter are, of course, worth a few pence, for they are used in felt making, but what is a few pence in comparison with six or seven shillings ?

(2) He must aim to have his rabbits in first-class condition and at the right age for killing some time between the beginning of November and the end of March, for only during that period are pelts in demand.

(3) He must produce good-quality pelts. What the trade wants is consignments of first-class skins with a good depth of fur, absolutely free from any signs of moulting and well matched. There are no buyers at anything like a decent price for mixed packages of inferior skins.

Provided one bears these essential factors in mind, would it not be good business to start and develop a rabbit farm on the lines of the present-day poultry farms ? some of you will ask. That is a question which can only be answered by experiment. To the best of my knowledge, no one has ever attempted anything quite so ambitious.

Yet I personally am optimistic enough to believe that *real* rabbit farms will come before we are very much older. Have not the Americans made big money out of fox farms ?

The publishers wish to thank Messrs. Watmoughs, Limited, of Idle, Bradford, proprietors of " Fur and Feather," for the use of seven photographs of breeds used to illustrate this book.

UTILITY
RABBIT KEEPING

CHAPTER I

HUTCHES AND RABBITRIES

THERE are two distinct methods of housing rabbits—the outdoor and indoor methods. If you ask which of the two is better, given equal care in each, I don't think I could answer. Some breeders swear by one, some by another, and both seem to meet with an equal measure of success.

The question of expense even is not the deciding factor. Outdoor hutches have to be constructed with more care than indoor hutches, and their extra cost has to be set off against the cost of putting up a shed.

It seems likely that rabbits reared in the open should be hardier than those reared under cover, but I have never actually proved that it is so.

OUTDOOR HUTCHES

Outdoor hutches vary so greatly that it is impossible to do more than describe a few common types. They may, in fact, be of almost any design provided they adhere to the one broad principle of being damp-proof and draught-proof.

The poorer rabbit-keeper will hesitate—and rightly so—to go in for an elaborate stock of hutches made of match-lining before he has found his feet. At the present price of timber a stack to house four rabbits might cost anything from twenty shillings to fifty shillings. Quite serviceable hutches can be made from several kinds of packing-cases at a total cost of not more than a few shillings, and these are quite good enough to make a start with.

Match-cases are as suitable as any. They can generally be obtained by ordering them in advance from the local stores. The cost varies according to the local demand, but usually the price will be from 2s. to 3s. each.

The wood is of a good thickness, and the size, 38 inches by 28 inches by 26 inches, allows the case to be easily adapted without much alteration. The lid comes in useful for making the division, and, being nearly the right size, requires very little alteration to make it fit.

To complete the hutch, the few extra materials required and the cost will probably be something like this :

	s.	d.
14 ft. of 1¾ in. by ⅝ in. battens (for making frame) ..	3	6
1 yard ½ in. mesh wire-netting (for netting frame) ..		6
1 board, 28 in. by 8 in. by ⅝ in. (for making door) ..	4	0
2 pairs of hinges		6
	8	6

The frame to take the wire-netting is made out of battens.

Another kind of hutch suitable for breeding stock is made from packing-cases used in the tobacco trade. These can be obtained fairly freely from tobacconists at a fairly low cost, and the hutch when finished will not have cost more than a few shillings to make.

I should always contrive to stand such hutches as these where they cannot get the full force of rains and winds, and I should endeavour to improve them before winter came along. It is impossible that a packing-case can remain watertight for long when exposed to all weathers. In any case they should not actually stand on the ground.

One way of keeping the hutches sound is to cover top, back, and both sides with roofing felt, and to fix a movable screen to cover three-parts of the wire-netted front. The latter, which can be made of canvas or timber, can be lowered when necessary to provide protection against driving rains. It may either be hinged to the hutch—which is preferable—or detachable.

An alternative plan is to build a rough sloping roof

of either felted boards of corrugated iron, projecting at least 2 feet over the hutches which, in this case, would then be stacked in tiers, the bottom one, be it remembered,

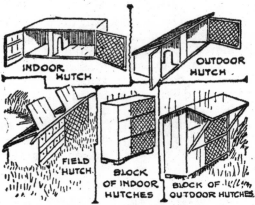

INDOOR HUTCH

OUTDOOR HUTCH

FIELD HUTCH

BLOCK OF INDOOR HUTCHES

BLOCK OF OUTDOOR HUTCHES

A selection of the rabbit hutches now in common use.

being raised a foot or so from the ground as a precaution against rising damp. When corrugated iron can be picked up cheaply this plan is an excellent one, and I have bred

BRANCH OF TREE

KEEP EDGES WELL SOAPED

Two schemes for those who find their stock are rather addicted to gnawing the wood-work of the hutches.

rabbits successfully through a most severe winter in hutches so roofed, though on very windy nights I took the precaution to drop a canvas screen over the whole front. This screen was on the roller-blind principle.

Another simpler plan for protecting stock kept in the open from inclement weather is to hang sacks over the front of the hutches, but not touching them by several inches.

A further suitable and more permanent method is to nail boards together, large enough to cover the whole front of the hutches, to which, with a pair of hinges, they should be fastened, to act as shutters. Such an arrangement is easily manipulated, and can be regulated to suit the conditions much better than sacks, inasmuch as a prop can be used to determine how far it is desirable to keep them open or closed. Indeed, a shutter of this description is a very useful permanent addition to any outdoor rabbit-keeper's equipment, as during the summer it can be used to protect stocks from heat, and also provide privacy for breeding does, besides being effective in keeping stray cats and dogs at bay at all times.

One rabbit-keeper I know, who was in a position to obtain a number of old barrels cheaply, constructed a most efficient rabbitry from them.

He first carefully removed the end in which is the spigot hole and thoroughly cleansed the barrel. Then, from slating batten (which is reasonable), he made a lattice floor to fit comfortably. The size, of course, was governed by the size of the barrel. Next, in the end of the barrel he had removed he cut a section large enough to allow the lattice floor to be removed cornerwise. This was done to facilitate cleaning.

The door proper was made of pieces of batten and was wire-netted. It was fitted over the hole in the barrel, to which it was fixed by means of hinges and fastened by a wooden button.

The base, roof, and uprights were made of rough packing-case timber, the roof being covered with a sheet of old tarpaulin freshly tarred.

HUTCHES FOR THE ADVANCED RABBIT-KEEPER

It will be remembered that I only recommend the rough-and-ready hutches described at the commencement of this chapter for the beginner. As you advance so your hutches should improve.

One of the best "advanced" hutches I have seen is shown in the drawings on page 17, which depict the top and bottom sections of a stack of four hutches, each 2 feet high. The total measurements of the stack are: height, 8 feet 2 inches; length, 5 feet 6 inches; and width, 2 feet.

This is the way it is made:

First take two lengths of wood, 8 feet 2 inches long and 2 inches square, and mark with your pencil places at 2 feet,

The safe and the unsafe types of hutch-door buttons. Rabbits are essentially mischievous and will worry a loose fastening day after day—with the inevitable result depicted.

4 feet, and 6 feet along the lengths; this will give the places where the floor of each hutch will come; 2 inches from the bottom of the lengths will be the bottom floor of all.

Now nail strips parallel between the two uprights at the bottom and top, and also between the points previously marked. Then nail a strip of the same dimensions as the two uprights from top to bottom at a point 15 inches from the left-hand side; this will divide the hutch into two parts, the smaller for the breeding compartment, and the larger for the wire run.

Perform this work over again for the back, making the following modification if the stack is to be placed against a wall or level surface : The strips which run across and form the supports for each floor should be let into the uprights so as to fit flush.

Now take two pieces of batten 3 feet 6 inches long, 2 inches wide, and 1 inch thick, and join the two frames together at the top, as indicated in Fig. 1 (M—N). Each piece has to project 1 foot 6 inches over the edge of the frame to prevent rain beating into the hutches. These help to form the frame for the roof. For joining the uprights at the bottom, two pieces 2 feet by 2 inches by 1 inch should be used. This finishes the inner frame of the stack.

Fill in with strips of 6-inch match-boarding the back, roof, sides, and floors. Also nail match-boarding across from D to E, leaving a hole in this partition for the rabbit to get through. If this hole is cut near the front of the hutch it will in some measure prevent the youngsters when born from falling out of the door, as the doe will put them as far away from this hole as possible.

Another way of preventing this calamity is to nail two small strips of wood 5 inches long on C and D, to form a slot for a board 15 inches by 5 inches to fit into ; thus when the door is opened the youngsters will not fall out. When cleaning out the hutch the board can be taken out.

All that now remains to finish the stack off is to fill in the front of the hutches. The door of the breeding part does not require a great deal of trouble to make. It can be fastened on with a pair of small hinges, and secured by means of a button.

Now to finish off the wire run. One of the favourite ways to fix the wire door is by hinges either at the top or side ; but there is a great drawback to this method. When the door is opened at feeding time, young stock of about a month old will, in their inquisitiveness, fall out unless extra care is taken.

This can be prevented by making a door as follows :

Take two strips of batten 4 feet long, 2 inches wide, and 1 inch thick, and two strips 1 foot 8 inches long by 2 inches wide by 1 inch thick, and make a frame for the wire with these. From the right-hand side of the frame so

By permission of   *"Fur and Feather."*

The Silver Grey

To face page 17.

made, measure off a point 9 inches from the end, top and bottom ; then nail a strip 1 foot 8 inches by 1 inch square, on the inside of the frame to join these points.

Another strip 9 inches long by 1 inch square should then be nailed across, from the side of the frame to the strip just previously put on, at a distance of 6 inches from the

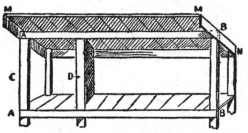

Fig. 1.—The topmost section of an advanced stack of hutches.

bottom batten in the frame. Thus you have an opening 1 foot 2 inches by 9 inches. In this should be fitted a wire frame to act as a small door.

To finish the stack off, waterproof felt can be nailed over the top.

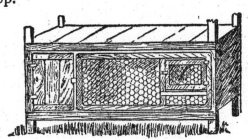

Fig. 2.—The lower section. Of course, as many intermediate sections may be included as desired.

Another advanced type of hutch is also depicted (p. 19). It contains many novel features, and is undoubtedly the last word in rabbitries. In the drawing measurements are given, but as it may not be practicable to use new wood, or sufficient boxes 2 feet square may not be procurable, it is

B

not suggested that these measurements need be exactly followed.

However, the uprights are important. These should extend above the top of the hutches so that a rainproof roof, projecting back and front, can be added, with a sufficient slope to carry off the water readily and keep the green-stuff, or anything else stored on the shelf formed by the hutch top, from the weather.

The remainder of the descriptive notes in the drawing are self-explanatory.

As the hay receptacle has stout galvanised iron wires meeting the floors at right angles, it is desirable that match-boarding be used for the floors (which may be well sprinkled with sawdust), running back and front, so that moisture shall not foul the hay. It is advisable when replenishing the hay " rack " to renew from the bottom, or mix the fresh with the old.

If boxes are used, arrange that the floor boards also run back and front, and that the floor of the upper hutch admits of no moisture dropping into the hutch below.

Another feature of this hutch is its provision of a canvas screen (suspended from C D). This will appeal to those who are compelled to place their hutches in an exposed position. The space below the hutches will serve to keep dry many odds and ends, or food kept in earthenware or other vermin-proof receptacles.

THE BEST POSITION FOR HUTCHES

The best position for outdoor hutches, single or in stacks, is against a wall or fence with protection from cold north and north-east wind. If they can face full south so much the better. But leave a few inches of space between hutch and wall to allow a current of air to pass round them.

Even then, sheltered as they are, they will need an occasional overhauling, especially before the commencement of the winter. Boards will have become warped by the sun, and must be drawn back into position by using screws. If the netting has loosened ever so slightly or if hinges and fastenings are not in good working order they should be seen to without fail.

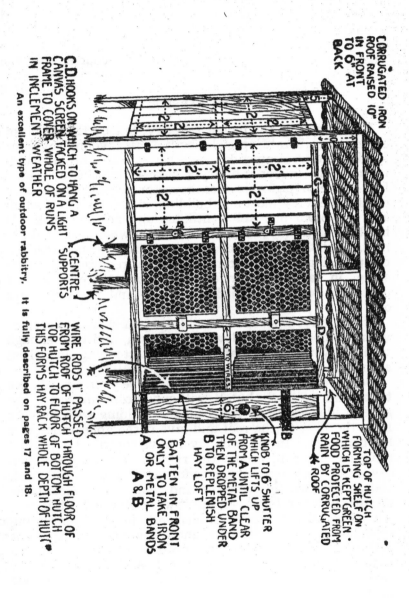

When a rabbitry is inclined to be draughty it is a good plan to give the hutches a coat of tar, then stick all over them sheets of brown paper and finish off with another coat of tar. This will keep them beautifully snug and warm.

Tar is also useful in making the .hutch floors watertight, a point that is of considerable importance when the hutches are stood in tiers, otherwise the urine soaks through and keeps the lower compartments constantly

The exterior—

—and Interior of a useful portable rabbitry.

damp. The tar should be well sanded and given ample time to dry.

Some fanciers cement the floors of their hutches but this is foolish practice, for cement is far too cold for the rabbit to sit upon, especially in winter. If tar is not fancied, apply a cheap, hard-drying varnish, carrying it two or three inches up the sides of the hutch.

PORTABLE HUTCHES

There is a third method of keeping rabbits which finds favour with a few—in portable hutches. There is a good deal to be said for the method, as the rabbits can be left in the open on all favourable occasions, and brought under cover whenever the weather is bad.

The hutch illustrated is as good as any I have seen. It provides quarters for two rabbits, and is constructed of packing-cases broken into pieces and rebuilt up. As the drawing shows both the interior and exterior of the hutch, there is no need for further description.   •

One final word before I close this chapter. Always use wire-netting of $\frac{1}{2}$-inch mesh. When larger meshed netting is used the rabbits can get their noses through and spoil the fur, a consideration in the case of a specimen destined to win show honours.

CHAPTER II

THE BEST UTILITY BREEDS AND HOW TO BUY THEM

A FEW short years back ninety-nine out of every hundred breeders would have told you that the only two breeds of rabbits suitable for utility purposes were the Flemish Giant and Belgian Hare. So they were, then; even to-day they are great favourites—as table rabbits—being hardy, well-fleshed, quick growers, and, most important of all, coming very close to that ideal which even now the buyer, probably unconsciously, still has before her, the Ostend rabbit.

But no longer have they the field all to themselves. Even when one is only considering the table-rabbit section there are several new breeds to be reckoned with on equal terms. When one also considers that newer utility rabbit branch, fur production, they are hopelessly outclassed. The pelts of the Flemish Giant and Belgian Hare are worth practically nothing to the furrier. Also there are breeds which, whilst possessing good table qualities, have also a first-class fur pelt.

Before setting out a full list of the latter we will refer briefly to the breeds which are table rabbits pure and simple, this for the benefit of those who, having a good market for table carcases, wish to confine themselves exclusively to this branch of the business:

Flemish Giant   Belgian Hare
Silver Grey   English
Japanese.

The Flemish Giant is steel grey in colour, closely resembling the wild rabbit, except so far as size is concerned.

It is a free breeder, usually having eight or nine youngsters in a litter.

The Belgian Hare is of a reddish colour and bears a striking resemblance to a hare. Its flesh is of fine quality, plentiful, and sweet. In so far as prolificacy is concerned it is in every way the equal of the Flemish.

The Silver Grey is more of a fancy rabbit than a table rabbit, but because its pelt used to appeal to the furrier, and because its flesh, though dark, was plentiful, it enjoyed a brief spell of popularity in the utility world—in the early days of pelt production when the advantages of keeping a dual-purpose breed first began to be thought about. Now it is hopelessly outclassed both as a table and a fur rabbit.

The English is, to my way of thinking, the best of all table breeds. Its size was at one time thought to be against it, but now it is realised that although it may not *look* as large as a Flemish, say, its small bones more than make up for that. Also it loses less than any other breed in the paunching operation, is a small eater, and fattens most readily.

The Japanese is a comparatively new breed which some think is shortly going to be a great favourite. It is curiously coloured, with broad bands of red and black, as will be seen on reference to the illustration facing page 120. It is a fine breed for those who are out to supply the demand for rabbits to be kept as pets but who would also have the table-rabbit trade open to them in case any youngsters may be left on their hands.

Now we come to the fur-bearing classes. Here they are:

Beveren	Chinchilla
Havana	Lilac
Argente de Champagne	Sitka.

The Beveren is far and away the most popular breed to-day, and is likely to remain so for a long time to come. Not only is its pelt ideally suited to the furriers' needs, but it has also excellent table qualities, almost equal to those of the Flemish Giant.

There are two varieties, the Blue and the White. The former is the greater favourite. It is really a most beautiful rabbit, being a clear intense shade of light lavender blue throughout, a colour that is in itself attractive to women. Even the eyes are a brilliant blue. The White is, of course, white in colour throughout, but in all other respects is similar to its confrère.

The Chinchilla's chief claim to favouritism is that its colour—slate blue at the base, pearl grey in the intermediate portions, merging into white and slightly tipped with black, with longer hairs of jet black interspersing—closely imitates the real Chinchilla which for so long has served for fur coats, stoles, and other items of Madame's wardrobe. The fact that it is one of the most expensive rabbits to start with is proof of its value. Chinchilla pelts never hang on one's hands.

The Chinchilla possesses good table qualities.

The Havana is another of those breeds whose pelts, being of a useful and popular *natural* colour, do not have to go through the dyers' hands. This latter is certainly an advantage when one recalls the " scares " that arise from time to time that dyed furs are injurious to the wearer.

The colour of this breed is a deep, rich brown in most cases, though it ranges through the shades from a dark coffee to a ruddy chocolate, the under-colour being a pearl grey. Breeders, however, are endeavouring to eliminate this under-colour on account of its obvious disadvantages to the furrier. The Havana is also a good table rabbit.

The Lilac, comparatively speaking a new breed, rivals the Beveren in beauty. Of an even pinky dove-colour throughout, its fur is wonderfully fine, soft, and dense, ideal from the fur man's point of view—so soon as enough pelts are produced to make him consider it more than he does at the time of writing. This breed has perhaps better table qualities than any other of the fur breeds, and, in addition, is a small eater and quick grower.

The Argente de Champagne is a great favourite in France, though only kept on a comparatively small scale in this country. Its colour is a beautiful shade of grey, approximating to old silver, with an undershade of slate blue.

The Sitka supplies the always great demand for a black fur, jet black in itself without the need for dyeing. Its colour is glossy jet, going well down into the fur, though the under-coat is dark blue.

PURE- OR CROSS-BRED

A few years back, cross-breeding was all the rage. In those days, though, we were endeavouring to produce the perfect table rabbit. To-day fur is the chief consideration, and, without the slightest doubt, you get the best fur on a pure-bred rabbit.

But does cross-breeding pay for table work? It *may* give a more meaty carcase, but no one has ever definitely proved it.

Pure-bred rabbits cost no more to keep, grow just as quickly, are very nearly, if not quite, as prolific, they are far more interesting to rear, and there is always the chance that something really good will turn up in a litter.

No, I honestly believe that breeding and keeping the varieties pure—always allowing for the fact that in isolated districts one may find an exceptional demand for the cross—will pay better than intermixing them, whilst a pure-bred rabbit will always command a better price when offered for sale than one whose parentage is not known.

Pure-bred rabbits are naturally going to cost you more, but which is the better, to lay out as much as you can afford on two or three good does, capable of producing quick-growing marketable youngsters, or on a dozen or so mongrels, which when fully matured weigh only 4 to 5 lbs. each, and from which it is almost impossible to breed youngsters which will scale 3 to 4 lbs. at twelve weeks old?

POINTS FOR PURCHASERS

But how can the beginner tell whether he is getting good stock when he is paying a good price? Here, I am afraid, is a difficulty. I have lately come across several unscrupulous breeders who misrepresent the age, pedigree, and relationship of the rabbits they wish to sell, and succeed in foisting upon their unsuspecting clients worthless rubbish which cannot but lead to failure or at least disappointment.

Of course, I do not altogether exonerate the purchasers

from blame. If they will try to buy a breed about which they have not troubled to learn a thing they are at least laying themselves open to fraud. And the moral is obvious. Before *you* buy learn as much as you can about the breed you fancy, and if possible get a friend to indicate the main points.

Next learn how to tell a rabbit's age, for a rabbit is useless for breeding purposes when it is more than three years old. I prefer always to buy my stock at from two to seven months old.

To the casual eye a young rabbit is much brisker than an old one, the eye is brighter, the teeth smaller and whiter. But the best test of all is the claws. In a young rabbit they are quite small and even and do not project beyond the fur of the foot ; in an older rabbit (eighteen months and over) they are thicker, curving, and sometimes project an inch beyond the fur.

Then you should learn the signs of health in rabbits. To bring a diseased rabbit into your hutches is fatal. Even when you are absolutely sure of the man who has sold you the stock it is advisable to isolate the new purchase for a period of not less than ten to fourteen days.

The signs of health are :

1. Ears, clean and free from wax.

2. Eyes, bright, full, and sparkling, the whites free from any yellow tinge.

3. General bearing, frisky and free from care, with plenty of life.

4. Good-tempered and showing no signs of pain when handled.

5. The droppings, firm and round, not soft.

Insist upon handling the rabbit. Pass your hand along its back and feel for any unnatural swelling. Look well down into the ears and well into the eyes, and make sure that there is no discharge from the nostrils. Then turn the rabbit up and look for sexual inflammation (see page 74) or vent disease and also examine both fore and hind feet for soreness.

Next get from the salesman a written guarantee that the buck you are buying is not related even remotely to the does you are buying, and that all are healthy. Also ask him whether they have been housed in indoor or outdoor hutches, and how they have been fed. Make a particular note of his method of feeding, as sudden changes of food invariably lead to trouble.

Be sure on all these points and you cannot go far wrong in your purchase.

The best age at which to buy a doe is eight months, when the weight should be 10 to 12 lbs. in the case of Flemish Giants, and 8 or 9 lbs. in the case of Belgian Hares. The buck should also be eight months old, but may be a little less weighty.

The best time to buy is about October, when breeders sell out their surplus stock on account of the scarcity of food. The rabbits will also have a chance to become well accustomed to their new quarters before they start breeding.

The best number to start with is three does and one buck.

The price you will be asked to pay for good stock will vary very much. Sometimes you can pick up a good three months old doe for 10s. 6d. ; sometimes you have to pay as much as £3 for a really good adult. It is never worth while paying less than £1 1s. for a buck.

BUYING THE BUCK

I am afraid the buck is often quite a secondary consideration with rabbit-keepers, who grudge the price they are usually asked to pay for him. This is folly in the extreme. He is just as important as the does, and it is worth paying a good price for him and lavishing just as much care on his choice.

Although your aim may not be to produce exhibition specimens, the mating to a good buck, as recommended, may produce a youngster that will prove to be worth ten times the amount paid for the parent.

Another important phase of the business is that the purchase of a pedigree stud buck is generally a good investment, as during the time you are satisfying your own requirements you can advertise his services at stud and give others a chance to improve their stocks, and incidentally add to your profits.

Summed up, my advice to the beginner is to—

1. Learn all you can about the breed you fancy.

2. Learn all you can about the breeder from whom you intend to buy.

3. Go in for a double-purpose breed unless, of course, you have an extra-special market for table carcasses.

4. Buy pure-bred does and mate them to a pure-bred buck. Pelts from cross-bred rabbits will never sell for more than a few pence.

5. Pay a good price for your stock to begin with. Far better content yourself with a couple of good does than half a dozen indifferent animals to be bought at the same price. The latter will earn you practically nothing ; the former will at least form the nucleus of a rabbitry which can safely be built up to any dimensions required.

6. Don't, having invested in a batch of scrub animals picked up at a " bargain " price and found they don't pay, go around spreading the tale that rabbits are " no good." You aren't in a position to judge !

CHAPTER III

FEEDING

MORE deaths among rabbits are caused by incorrect feeding than by sickness and disease. Yet there is no live stock that is less trouble to feed.. They will eat almost anything at almost any time. Trouble only comes because insufficient care is given to the condition of the food or its preparation, because of gross irregularity in the feeding times, or because the diet is changed suddenly from one thing to another without any thought.

Let me take these feeding " don'ts " one by one.

1. *Don't feed mouldy, sour, dirty, or partially decayed meals, garden or house refuse or oats, or frosted or very wet green food.*

Mouldy, sour, or partially decayed food will in all probability be refused by the rabbits unless they are exceptionally hungry. If they do partake of it, internal troubles are almost bound to arise.

For this reason only just as much food as the rabbits will clear up at a time should be given them. Food not required and left in the hutch soon becomes contaminated or goes sour. Then perhaps one of the rabbits, feeling peckish, nibbles at it and trouble follows. You can soon gauge a rabbit's appetite if you attend to it entirely yourself, and once that is done you will find very little variation.

The caution against feeding frosted food is just as imperative. The strongest of rabbits cannot eat frosted foods without very serious and more often fatal results. Don't gather green stuff or dig roots in frosty weather, therefore, or if you must, take proper precautions to get the frost out of the food before the rabbits have it.

In the case of roots, bring them into the kitchen and leave them close to the kitchen fire *all night*. Don't put them right on top of the stove, but just close enough to thaw the frost. Green stuff that is badly frozen should also have an hour or so in front of the fire, but if only slightly frosted you may give it a rinsing in *cold* water, followed by a good shaking or, better still, taking a bunch in your hands, swing it about lustily a few times to remove the excessive moisture. This should also be done with food which is gathered when very wet.

Food can only be kept clean if fed in troughs. There are many types of troughs, two being shown on page 33.

2. *Don't feed at irregular hours.*

This leads to the question : How many times a day should a rabbit be fed ? Nearly all the utility breeders of note feed three times a day, and if you can conveniently manage it you should do the same ; but if you are away all day, rather than leave the feeding to some untrustworthy person, feed only twice a day.

On the three-meal system the best hours are 8 a.m., 12 noon, and 6 p.m. ; on the two-meal system, 8 a.m. and 6 p.m. These hours may be earlier or later to suit your own convenience, provided you stick to one definite arrangement, and do not make too long an interval between meals.

Keep rigidly either to the two-meal or three-meal system, whichever you choose, and to the hours, be they what they may.

3. *Don't follow any fixed feeding rule.*

(To this I add the reservation except so far as concerns the hours of feeding and the number of meals a day.)

To begin with, a rabbit appreciates a change of diet just as much as you or I do and if constantly fed on the same food will begin to go downhill fast. Then it must be borne in mind that rabbits, like all other animals, differ constitutionally, so that it is quite impossible to lay down a definite feeding plan to be followed in every rabbitry.

Some time ago I carried on experiments with two rabbits with dandelions as the chief food, and oats and wheat in moderation. At the end of six weeks rabbit No. 1 lost flesh, became scraggy in coat and was generally listless. No 2. however, carried a substantial increase of flesh and

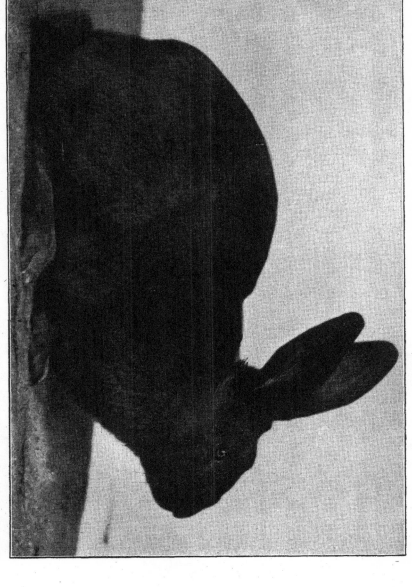

The Havana, a beautiful rabbit whose pelt is in
much demand.

To face page 35

was in every respect the finer rabbit of the pair either for market or breeding.

In experiments later on with chicory, hedge parsley, and lawn grass clippings, the result, although not so pronounced, was nearly so.

The only thing is to consider the general principles of feeding as a guide and to add or substract to them as circumstances require. It is no use, for instance, con tinually to force meal mashes on to stock that do not relish them. If this is done one's object, which is to obtain early

The best type of feeding trough to use.
It is made of earthenware.

A useful type of home-made trough. A metal flange prevents waste of food and gnawing of the wood.

development, is defeated rather than promoted, no matter how good are the ingredients of which the mashes are composed. A handful of green food, plucked at no cost, and which is eaten with avidity, will much better serve the purpose.

The true guide must be the way in which the rabbit accepts the food given to it. If the hutch is littered with different kinds of food from one meal to another, the food is not serving a useful purpose, because, obviously, it is not appreciated.

No doubt a great many of the " fancies " a rabbit has

C

for a particular kind of food are due to the manner in which it was fed after being weaned. When these youngsters are sold, scattered in various rabbitries all over the country, and consequently introduced to new and strange foods, their likes and dislikes evidence themselves.

That is why, as I have already pointed out, it is always wise when buying fresh stock, to ascertain how they have been fed, and also how many meals a day they have been accustomed to, so that the same principles may be continued. Even if the food given is eaten, different feeding methods will often cause a check in the growth of young rabbits, and this is a point against which we must be particularly guarded.

4. *Don't indulge in sudden changes of diet.*

A sudden jump from one food to another has led to the death of many a good rabbit, and it is a mistake that all of us have made at one time or another. As the different seasons come and go so some foods come in and others go out, and it is impossible to stick to any one particular class of food for long at a time. But the supplies do not cease abruptly. The change from one to another can always be made gradually.

Consider for one moment the vast difference between succulent green foods and dry oats and bran ; yet such a change is often practised, and, true, sometimes successfully accomplished. Far better, however, to alter the diet gradually.

From early summer until late autumn, many stocks of rabbits practically live entirely on green foods. Then, with the advent of winter, this is all changed. But the best way to go about it is not to drop the succulent foods in a day, or even a week, but gradually to decrease the supplies of green food as it becomes appreciably shorter, so that by giving a little each day until the supply is finally exhausted the change becomes a gradual one, and does not interfere with the constitution of the stock.

A small piece of swede, carrot, or beetroot will help to keep the rabbits in good health whilst they are being " weaned," as it were, from the natural wild herbs they have so long enjoyed.

5. *Don't buy foods in small quantities.*

This " don't " affects the rabbit-keeper rather than the rabbit, but it is equally as important as the others. If he buys in small quantities his profits are going to be half what they should be ; if he clubs with several neighbours and buys in bulk he is going to save 30%, 50%, and sometimes even 100% in the purchase price, besides getting better food.

I know personally several men who are to-day paying perfectly absurd prices per pound for oats, quite oblivious of the ordinary market price of good, sound oats at the bushel or the quarter.

Co-operation will be the means of securing the corn at far cheaper rates. Co-operate, therefore, with others in your district and buy by the quarter, not by the gallon or peck, the pound or the stone.

Coming now to the various types of food suitable for rabbits, wild green food is first on the list, for not only is it the most economical to use, but it is the natural food of rabbits, and therefore they thrive on it better than on any other.

GREEN FOOD

Green foods are plentiful from early spring well into October, and provided you exercise due care in starting with it, that is, start with a small amount and gradually increase the allowance, you may feed them freely. But remember that rule as to variety. Keep on changing about from one thing to another, for all can be had for the gathering.

The question, " How much green food may the stock really have ? " naturally arises, but is soon answered. So long as all you give them is cleared up and they seem to be asking for more, there need be no hesitation in giving a little more, but if any is left in the hutch, cut down the allowance. If green food becomes soiled and is then eaten, trouble in the way of sickness is likely to ensue.

The truest indication as to the correct quantity to give is the state of the appetite of the consumer, and as rabbits generally satisfy themselves in this direction immediately their food is given to them, it is a very easy matter to decide in every case whether or not sufficient has been apportioned to them. If they do not want their food they are quite indifferent to your attentions, whilst those that are hungry generally make a good meal and finish with it.

You may notice that there is a slight looseness of the bowels when feeding heavily on green food. This can be counteracted by a regular supply of hay or hay substitutes, which, in fact, should always form part of the daily ration. Oats, too, or oat substitutes, should also be included in the daily diet, as they keep the flesh firm and add weight.

Coltsfoot.

WILD GREEN FOODS

The best sorts of wild green food are as follows. Most of them grow plentifully in every district.

Blackheads. This plant is similar to those of the plantain family. It has very long stalks, at the end of which is a hard black ball containing the seed. It grows largely in hedgerows and on the banks of fields.

Coltsfoot. Is found growing luxuriantly on railway embankments, coal-pit mounds, in sand-pits, and on other waste land. It is an excellent food and thoroughly enjoyed if given with other green food. The plant grows in clusters with a dwarf habit.

Dandelions. Are too well known to need description. At one time they were practically the only wild green food given to rabbits, and even to-day they are greatly favoured. They have also a good tonic effect on the stock.

Groundsel. A very common weed possessing good food value.

Hedge Parsley. This is the earliest and most useful of

Hedge Parsley.

Hemlock.

all wild green foods. It has fern-like leaves and triangular-shaped hollow stalks, and grows in clusters on the banks of hedges bordering fields. Most lanes also provide a supply if a careful search is made for it.

On no account confuse hedge parsley with hemlock, which is deadly poison, and has carried off whole stocks of rabbits in one night. Hemlock is a dwarfer plant and has a coarser, broader leaf, a round stem with small purple spots, which often run up into the leaves, and a very disagreeable smell when crushed.

Marsh mallow. This is looked upon by some as solely the plant for herbalists' use. Rabbits relish it, however, and it is a good tonic for them. It may be recognised by its pale blue flowers, which

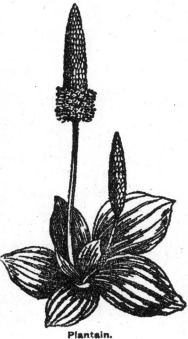

Plantain.

grow in clusters, and its leaves, which resemble those of a geranium.

Plantains. There are two or three varieties of plantain some growing tall and some of a dwarf habit. They are easily recognised by their long seed stalks, which are known in many parts as " rats' tails," and sold as food for such birds as the bullfinch and linnet. Plantains are a good food, but are not used so frequently as they might be, on account, I believe, of their value being unknown.

Shepherd's Purse.   Marsh Mallow.

Shepherd's purse. Is an excellent plant, having a tendency to check looseness of the bowels. It grows mostly at the base of walls that border footpaths.

Tree shoots. Sprigs and leaves of hawthorn, blackberry, elm, and oak are good as an occasional food, but should not be given very often.

" GARDEN WASTE " FOODS

The owner of a kitchen garden or allotment has no need to go into the fields and along the hedgerows to find food for his rabbits. He has a plentiful and costless supply at home practically the whole year round, which if not fed to stock merely goes to waste.

The following list shows what a great variety of foods can be obtained from the average garden :

Lawn mowings.	Pea haulm.
Bean haulm.	Cauliflower leaves.
Carrot tops.	Turnip tops.
Pea pods.	Outer Lettuce leaves.
Radish tops.	Mint, Sage, and Thyme.
Nasturtium leaves.	Sunflower leaves and seeds.
Maize foliage.	Strawberry runners.
Cabbage leaves.	Beet leaves.
etc.	etc.

In addition to the leaves of cabbage, cauliflower, etc., the stalks are also useful if they are split up. The rabbits eat out the core and trim off the outsides, leaving but the inner skin. Cabbage leaves, by the way, should only be fed sparingly unless plenty of hay is given with them, as without hay they are liable to cause scours and also cause a rather unpleasant odour in the hutch.

It will be noticed that sunflower seeds and leaves are included in my list, and this may surprise many rabbit-keepers. Few seem to realise that they are an excellent food for adult stock, the seeds being especially valuable for does with suckling young. They contain a large percentage of oil, however, and therefore should not be given too generously. A dozen seeds every other day is usually sufficient.

The seeds should be crushed or bruised and mixed with the meal mashes. If they are fed separately some rabbits will refuse to eat them.

A CROP TO GROW ESPECIALLY FOR RABBITS

One of, if not the, most valuable green foods is chicory. It is full of nourishment, and I have never known a rabbit refuse to partake of it. It is most easy to grow, thriving in any odd corner, and once planted will multiply rapidly year by year. It is, therefore, an ideal crop to grow for rabbits when other green foods are difficult to obtain. To town rabbit-keepers it should be particularly valuable.

There are two ways of starting a chicory bed, by sowing seed or by planting one-year-old roots The latter is the better, because a supply of green food is available almost immediately after the bed is planted up.

The roots are planted deeply two or three inches apart in early spring, and allowed to make a fair growth before cutting commences. The seed is sown about the end of June in shallow drills, and the plants thinned out to three inches apart when they are large enough to handle.

ROOTS FOR RABBITS

When the supply of green food ceases, or almost ceases, in the autumn or early winter, some substitutes have to be found, for the adoption of a perfectly dry system of feeding—hay, oats, bran, etc.—would quickly lead to trouble apart from the expense entailed.

Fortunately we have in various roots as good substitutes as we can reasonably want. In changing from green food to roots, let the same warning as to the gradual introduction of the new food be remembered. A sudden change must on no account be made. Neither should roots form the whole dietary. The same amount of hay or hay substitutes and oats or oat substitutes will be necessary as in the case of green foods.

All the following roots may be fed with safety, though a particular rabbit may not care for some varieties mentioned. When this happens the animal should not be forced to eat what it does not care for, but should be tried with a different variety :

Beetroot.	Potatoes.	Parsnips.	Carrots.
Mangels.	Swedes.	Kohlrabi.	Turnips.

Beet, parsnips, potatoes, carrots, and turnips will be found in almost every garden or allotment, though I do not suggest that the best of them should be fed to the rabbits. All the stock should have as a rule are those which are small, slightly diseased or otherwise unfit for use or sale. In the case of diseased roots, cut away the bad portions.

Some breeders make a point of sowing a bed of carrots specially for their stock, and this is not at all a bad idea when ground is available. It is always cheaper to grow than to buy. Other breeders visit all surrounding allotments and buy up for a few pence all roots that are not fit for culinary purposes. This also is a good idea, especially as the roots can be stored in clamps until required for use.

To clamp them, throw them up into a conical heap in a dry place, cover them with a thick layer of straw, then

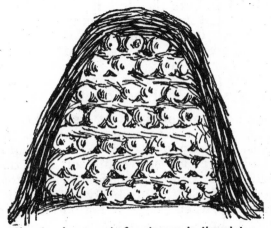

How to clamp roots for storage in the winter. The covering must be thick enough to keep out frost.

a thick layer of earth, leaving one or two openings in the earth layer at the top of the heap for air to get to the contents. These holes must be well covered up in frosty weather.

Carrots are the best of all roots for rabbits. Occasionally one comes across a rabbit that has no particular liking even for them, but such are isolated cases. Being very rich in heat-producing matter carrots are particularly suitable for winter use, when the body requires warmth. Such a food at that season is therefore very desirable, indeed necessary. The carrots should be cut into 2-inch lengths before being fed.

Swedes rank next in order and seem to be enjoyed by all rabbits. They withstand frost far better than most other root crops, and do not wither so quickly after they have been dug up. They must be cleaned before they are fed, but need not be peeled ; the rabbits will do that. When starting with them give each rabbit a piece the size of an egg to get them accustomed to it, and then gradually increase to the required quantity.

Mangels, too, are enjoyed by most rabbits but are not so frost-proof and are therefore not so easily stored. On no account should they be given to the stock until well into the new year. When clamped for some time the acids they contain are chemically converted into sugar and not until then are they safe.

Incidentally, the leaves are a most nutritious food, and should always be used if they can be picked up cheaply.

Potatoes are sometimes a very useful food, although I admit their prices of late have been prohibitive, and most people have rightly hesitated to give them to rabbits. They should be boiled until quite soft, and then dried off with meal. They can be given four or five times a week.

The only trouble that is likely to arise from feeding boiled potatoes, whether diseased or not, is a form of windy colic, which can easily be avoided by only feeding them when warm. In the majority of cases of colic arising from the feeding of potatoes, it will be found that more than the animals have been able to consume at one meal has been given. A portion has been left before them until it has become cold, the rabbits on regaining their appetite later have eaten the cold mash, and in consequence get indigestion.

Beetroot (uncooked) is always acceptable, and sometimes small roots unsuitable for culinary purposes can be picked up for a mere song. Their only disadvantage is that they will not keep for long.

Turnips are eaten with relish only by a few rabbits; some will not touch them at all. They are, too, very watery, and care should, as a consequence, be exercised not to give too much at a meal.

Kohlrabi and Parsnips will not be appreciated by all rabbits, but when they are may be fed once or twice a week to make a break in the usual routine.

Apples. Although perhaps hardly coming under the category of " roots," it may interest many to know that as a change I have frequently given a piece of apple to my rabbits during the winter months and find they appreciate it greatly.

DRY FOODS

A certain amount of dry food must always be fed in conjunction with green foods or roots. The best dry foods for rabbits are hay, and oats, and barley, but to use these in any quantity and regularly would be too expensive for most of us. There are, happily, further dry foodstuffs, enumerated below, which may be worked in judiciously with them :

Oats are reckoned to be the best of all dry foods for rabbits, and when the price is right this is undoubtedly so. When you are asked to pay the tremendous prices that are now being asked for oats of almost any quality, however, you are inclined to be content with the proverbial " just as good."

The idea is prevalent among some breeders that the crushed or " kibbled " kind are not only the most desirable, but also most economical. With this, however, I cannot agree. I have used both the crushed and the " whole " varieties in large quantities, and am certainly convinced that the latter are, in the end, the most beneficial, and also go farther.

When young rabbits are first put on hard food the crushed variety are better for them, because they are much more suitable for small teeth. This, however, is all I have found in favour of this kind of oat. True, rabbits will sometimes " shell " their oats, but such are—at any rate with me—the exception and not the rule.

A few " whole " oats in the menu for the stock two or three times a week can always be given with advantage, and unless the price is prohibitive this should be managed.

Wheat. It is safe to assert that very many stocks of domesticated rabbits have rarely, if ever, tasted this grain, and few of us need wonder at this when we remember the prohibitive price and universal scarcity of wheat in these after-war days.

Containing as it does something like 12% of flesh-forming matter, and 70% of starch, it is very stimulating. It should not be used, when obtainable, so largely as oats, however, a very good way to serve it up being one part of wheat to two parts of oats.

I referred above to rabbits " shelling " their oats—that is, they crack the grain, from which they extract the kernel, and leave the shell. The mixing of wheat and oats in the proportions named very often causes a cessation of this habit, and those who give whole oats occasionally should if possible mix a little wheat with them both from the point of view of it being a beneficial addition to the feeding list, and also as a means of bringing about a cure for the shelled oats habit.

Barley can be used, when obtainable, the same as wheat, but although analysis proves it to be a grain of great food value, it is not relished nearly so much as wheat, and, surprising as it may seem, some rabbits will not touch it unless they are very hungry.

Bran. At one time bran was so popular as to be considered the last word in dry food for rabbits. Times change, however, and bran has lost much of its popularity. It has very little nutritive value, particularly so when given in its dry state, and it is probably due to the recognition of its extreme dryness that we hear of tea-leaves being used along with it for the sake of providing moisture, and thus aiding digestion.

When used in any quantity, the greatest benefit is derived by mixing it into mashes with other ingredients. (*See* p. 47.)

When you buy bran you should choose that composed of large flakes, and which when rubbed in the hand leaves behind a residue of flour.

Hay. The value of hay as a food for rabbits is very great. Bulky food is required for distending the stomach, which cannot properly perform its duties without, and when some rabbits get a rich, condensed diet, oats, for instance, and are short of green food or roots, as sometimes happens in the winter-time, a large handful of hay, either upland or meadow, will be greatly appreciated, and give the animals that comfortable feeling of fullness.

Hay, again, will help to balance an excess of green food, or roots, and where it can be obtained should be a regular article of diet in any rabbitry. It will also fatten up stock in very quick time.

The kind of hay used does not matter a jot provided it is sweet and clean. On no account use dirty or dusty hay; it is only left untouched, or is regarded by the rabbits merely as bedding. A few extra coppers are well spent in this direction. I do not profess to be a judge of hay, but personally I choose that with plenty of herbage in it. My own stock certainly appreciate that kind, and do well on it.

Many rabbit-keepers have not the accommodation to stock a large quantity, and perhaps it has not occurred to them that most hay and straw dealers will let them have a bagful for a comparatively moderate figure. It must be remembered, however, that hay is high priced in these days.

I have a bag (a sugar bag, costing twopence at my grocer's) which I take a day before I require the hay, and get it packed for a shilling. This, bear in mind, is the best hay for eating and not sweepings solely for bedding purposes. It is the loose hay that remains when trusses are broken up.

Straw. The straw of most cereals is eaten readily, and is a good, if bulky, food. It is not advisable to feed it to rabbits under eight weeks old, however.

Lucerne, trifolium, tares, rye-grass, and clover are also excellent rabbit foods.

OTHER FOODS

Acorns are a much neglected but none the less valuable food. An analysis of their food constituents shows albuminoids, 2·5; fats, 1·6; carbo-hydrates, 34·5; husk 4·4; mineral salts, 1·0; water, 56·0. Albuminoids promote the growth of lean flesh, fats give heat, carbo-hydrates are mainly fat-forming, so that acorns should prove a good winter and fattening food for rabbits. They must, however, be used in moderation, say, once a week or three times a fortnight at most, as they are rather astringent. They are best crushed with a mallet and mixed with bran before being fed.

Malt culms. To those rabbit-keepers who have a difficulty in getting a regular supply of roots or green food for their stock in winter, a sack of malt culms will come in very handy. They are the sprouts from the barley, grown in the process of malting, and rubbed off the grains ; they are supplied in a perfectly dry state, and if kept so, will be good for a very long time ; being extremely light when dry, a sack of six stone, costing a few shillings, will provide a very large quantity of one of the best substitutes for green food. Water has to be added and the culms allowed to swell ; use hot water for preference, and cover up the receptacle to keep in the steam ; the aroma given off in the proqess is not displeasing.

Brewers' grains. Brewers' grains, which in a wet state can be obtained in normal times at some breweries very cheaply indeed, form another cheap food for rabbits. The ale, or light-coloured, are the best liked, and if given warm and fresh most rabbits will eat them alone with evident relish ; after they have cooled a small proportion of middlings mixed with them will be found necessary.

It may even be necessary to dry off fresh grains with meal before the rabbits will eat them. When stock persistently refuse them even then the only way is to feed them when the stock are particularly hungry, preferably as the first feed in the morning. Once they have become accustomed to the flavour, the rabbits will devour the grains greedily.

In those lucky districts where they are procurable the breeder should, if possible, arrange to get fresh supplies weekly, as, owing to their perishable nature, they are apt, especially during warm weather, to turn sour. A good way of storing them is to press them down tightly in barrels, in the bottom of which a few holes have previously been bored so as to allow the liquor to drain away. As each layer of grains is pressed down it should be lightly sprinkled with salt. When packed this way, the grains will keep sweet for about a couple of weeks.

Bread. This is not now, of course, used nearly so much as it was. Nevertheless, every household has its crumbs and stale pieces, which are still not made the best use of. This is a great pity, as they can be used up either in their ordinary condition, toasted, or scalded, the water being drained off, and a little milk added.

Another excellent way I have adopted is to make them crisp by heat, and then, with a flat-iron or a rolling-pin, smash them up into pieces about as large as peas, and give them mixed with oats. They are relished in this way, and it is my favourite method of using up our dry bread-crusts.

Acorns are an excellent rabbit food if crushed
and mixed with bran.

Do not use new bread, and always avoid any which happens to have gone mouldy, or which has been buttered.

Coarse oatmeal is useful as a change, but is best used as a base for a meal mash.

Linseed is an excellent conditioner for rabbits intended for exhibition, giving the coat a fine sheen. It should be steeped in water and then well drained.

Barley meal and buckwheat may both be used very sparingly.

Meal mashes. Mashes made of boiled potatoes and peelings, brewers' grains, barley meal and bran or middlings and bran are a fine substitute for oats.

The mashes should not be made too moist. Add boiling water until the mash is of a soft crumbly consistency and the particles adhere together when squeezed. It is unwise to give too large a quantity at one feed, as any that is left over quickly goes cold and is then rarely eaten. A piece the size of a cricket ball will usually be found sufficient.

FOR TOWN RABBIT-KEEPERS

You very frequently hear that it doesn't pay to keep rabbits when living in a house with no garden and when wild green food is not available. This is ridiculous. There are plenty of town rabbit-keepers. One of our most successful breeders lives in Peckham, and his rabbits never taste wild food from one year's end to another, and hundreds of other breeders are similarly placed. Yet all find their rabbits pay them well.

Says an authority: " The townsmen who say rabbits don't pay are always those who are unsuccessful, and who think they know more about the breeding and management of rabbits than any one else. If you take the trouble to inquire of them as to how and where they purchase food-stuffs you will learn that they buy, not co-operatively, but in small quantities, taking, say, two or three quarts of oats and a few pennyworth of hay twice a week from some local corn chandler. If they only went into matters they would find that they are paying very dearly for everything they use.

" Of course, the town man is not in the same position as the country breeder so far as the collecting of wild green food is concerned, but we have greengrocers in every district from whom we can, for the greater part of the year, get a good supply of greenstuff, usually merely for the asking and the fetching. I have usually been able to get a grand lot of cauliflower leaves for my stock for nothing, and the greengrocer from whom I get them tells me that he is only too pleased to have them out of his way."

I also at one time in my career have found cauliflower leaves most useful, but being uncertain of my supply I had to find means to store them. Of course, leaves of all kinds go yellow and lose their usefulness as a food within

a week unless proper precautions are taken, but a breeder suggested in *The Smallholder* an excellent method of arresting the decay :

" I spread a quantity out flat in a cube sugar box," he told us, " and sprinkled them all over with salt. On the top of this I put another layer of leaves, then another layer of salt, and so on until the box was full. At the end of three weeks, I took off the weights I had used to press the contents of the box closely together, and found, to my delight, that many of the leaves, principally those at the bottom of the box, were quite usable.

" A certain number had withered and seemed to have affected the next layer to them, but to me the result, nevertheless, brings a great amount of satisfaction. I feel sure that there is only some particular feature to find out to render such an undertaking a success.

" Before using the leaves I carefully wiped the salt from them. I fed them to adult rabbits only, and they were regarded as an undoubted treat."

Other foods available to the town rabbit-keeper are :

Potato peelings.	Bread.
Small potatoes.	Carrot tops.
Broccoli leaves.	Straw used for packing.
Tree-leaves.	Pea and Bean shucks.

I have already described the way to feed potatoes.

Potato peelings are prepared in somewhat the same way. Wash them, boil them until soft, mix them with equal parts coarse middlings and barley meal into a sticky mash and give to each rabbit a piece as big as a teacup.

Leaves. A large quantity of leaves can be gathered in public parks and enough can be stored for use during the entire winter. The leaves of the yew, privet, and, in fact, most evergreens are poisonous and should be avoided.

When storing leaves intended for feeding, tread them down tightly in boxes, and sprinkle them with a little salt ; where they are only wanted for bedding during cold weather, pack them tightly in bags, and store them away off the ground to prevent the drawing of dampness.

Whichever method of storing is adopted care should be taken to see that the leaves are perfectly dry and clean

D

before they are packed ; otherwise they will soon become heated, turn mouldy, and give off a musty smell.

Packing straw. The straw referred to is the oat straw used for packing purposes. If a few inquiries are made at local shops it will be found that in many instances a supply can be obtained for a very trifling amount. In some cases the shopkeeper will be only too pleased to give it away in order to keep his premises clear.

This straw should only be fed to rabbits over nine weeks old.

Carrot tops. If your greengrocer will not let you have carrot tops at a nominal sum, you should grow them your-self—the tops, *not* the carrots. Hollow out the thick parts (the shoulders) of the carrots you have used in the kitchen and place them in shallow vessels full of water. Keep them indoors in the best light obtainable and the rapidity of growth of the pretty foliage will astonish you. See to it, however, that the vessel is never allowed to get dry.

I have on several occasions been able by this method to get a good handful of green food in the depth of winter when other greenery was exceptionally scarce.

SOME MODEL RATIONS

It is very difficult indeed to give model rations, as feeding depends entirely on what the breeder has by him or can obtain most conveniently. The rations given below should, therefore, be considered merely as a guide.

1. *For young growing stock.*

Morning. Green food, with a small supply of either dried grass, hay, or oat straw.

Midday. Green food, with perhaps a lump of stale bread.

Evening. Green food, with a handful of hay or oat straw, and a piece of mash, or a handful of crushed oats.

2. *For adult rabbits.*

Morning and midday as above, only in greater quantity.

Evening. Green food, plenty of hay and two handfuls of oats.

3. *For breeding does before kindling.*

A doe intended for breeding purposes does not require foods that are either forcing or of a fattening nature. Not only are fattening foods usually the most expensive, but, on the other hand, if the doe is in an overfat condition the matings will often prove to be ineffective. On no account is it advisable to keep to a fixed diet, as rabbits like a variety, and unless a change is given they will lose their appetites and soon commence to lose flesh, too.

The following ration will act as a useful guide :—

Morning. Handful of meadow or clover hay; good supply of green food.

Midday. Green food.

Evening. Small handful of oats, or an occasional feed of meal mash, green food and hay.

How to store leaves for feeding to rabbits.

4. *For suckling does.*

In order to provide sufficient milk the doe requires to be fed generously and in variety. A good supply of green food, given two or three times daily, will help to promote a heavy flow of milk. To improve the *quality* of the milk some food that is more concentrated, say one of the patent rabbit foods or a mash made of equal parts barley meal, middlings, ground oats, and bran, should be given once a day.

Sweet, herby meadow hay ought always to be before her. It is a good food, and it is surprising at times the amount a doe will consume when she is suckling young. I have found it excellent in counteracting that looseness of the bowels which green food often causes when fed in large quantities.

Other varieties of hay, such as lucerne and clover, can be given as a change. They are greatly relished by the stock, and where they can be purchased at a price within the reach of an ordinary pocket, they are a reliable and at the same time, good food.

Oats can be given when procurable three or four times a week. They should be sweet in smell. I do not advise crushed or bruised oats, because in the majority of cases, the crushed oats offered for sale in the retail shops are of inferior quality.

Here is a sample ration when you can obtain wheat :—

Morning. 2 parts oats and 1 part wheat, and on alternate days a mash of equal parts barley meal, middlings, and best broad bran ; green food and hay in quantity.

Midday. Green food and hay.

Evening. As in the morning.

5. *For stud bucks.*

There is hardly anything better for getting and keeping a buck in condition than a variety of foodstuffs, especially green food, which is plentiful at one time of the year. In addition good sweet herby meadow hay, and first cut clover hay, given alternately, has an excellent effect.

Other suitable feeding stuffs are good 40-lb. oats, milling offal, etc., while as a pick-me-up a handful of maple peas that have been soaked and allowed to sprout, can be given two or three times a week.

6. *For fattening rabbits.*

Fattening rabbits do not need nearly so much green food, but must have a largely increased allowance of hay and meals mixed to a crumbly, moist consistency with skimmed milk, or, if that is not obtainable, water. Boiled potatoes, soaked bread, etc., are all of a fattening nature, while a piece of oil cake an inch square daily will put a fine finishing touch to the rabbit.

A good ration would be :—

Morning. Boiled potatoes dried off with middlings, green food and hay.

Midday. Green food, hay, and a few oats.

Evening. Equal parts barley meal and middlings, or boiled rice ; green food and a good supply of hay.

CHAPTER IV

THE WHOLE ART OF BREEDING

As I have said in a previous chapter, an excellent start may be made with three does and one buck. But make sure that you *have* three does and one buck, not two does and two bucks. Many an inexperienced rabbit-keeper has attempted to " mate " two bucks, with disastrous results.

You can generally tell the sex at a glance, the buck having a much broader and shorter head than the doe. If this does not convince you, turn the rabbit on its back and examine it for sexual difference. This may puzzle you a little at first, but it is impossible here to go into further details and you will soon become expert.

When experience is gained breeding may be carried on throughout the year, but the best time for the beginner to start operations is in the spring, because this is the natural mating season of rabbits.

Before they are mated, both the buck and the doe must be in prime condition. The doe should be hard and firm, and at the same time should not in any way be overfat. Further, the coat should be in good condition, and not loose or moulty, and, above all, the animal should be free from cold or any trace of disease.

When a doe is mated in a poor or unhealthy state, the chances are that the offspring from such a mating will prove to be poor and weakly ; it is this kind of litter that is always unprofitable. A doe that is strong and vigorous is the one that produces the quick-maturing youngster, ready for marketing at an early age. It is always possible to put back a doe for a few days or until she is in full health again before she goes to the buck. If she does not seem to pick up properly you will find it far better and cheaper in the long run to kill her off.

DON'T NEGLECT THE BUCK

I am afraid many breeders are often given to neglecting the state of health of the buck, even if they see that the does are perfectly fit. What a foolish policy this is ! The buck has just as much influence on the litter as the doe, and if he is weak or ill or poorly nourished these troubles are almost bound to crop up in some or all of the litter.

Yet in many instances he is given some dark corner in a small hutch and denied various tit-bits simply because he has had the misfortune to be born a male. He should fare equally well in the matter of food and attention as any other inmate of the rabbitry. Whilst he is yet young he should have particular attention to ensure that he grows as large as possible, for the bigger he is the more likely are his progeny to be large.

When run with the doe his coat should be in the pink of perfection (a few minutes' stroking with the bare hand each day will ensure this), his flesh hard and firm, his eye bright, and his general bearing merry and bright. If he seems at all out of sorts he should not be used until he has completely recovered.

While, as I said, many breeders neglect their buck, others go to the other extreme and allow him to get so fat that he is unable to perform his duties efficiently. In many complaints I have investigated in which the breeders declared their does had accepted service but had not kindled, I have found that the bucks alone were at fault ; being excessively fat. They were dull and sleepy, and did not appear even to have any inclination to carry out their duties.

If the stud buck is allowed to get into this condition, he will, in a large number of his matings, prove to be sterile.

Of course, it is easy enough to reduce the surplus fat in about two or three weeks by regulating the feeding, but in doing so you are bringing the buck's health down into a low condition, and before he is again really fit for service you will have wasted from five to six weeks. This means a serious loss. In addition to the cost of keeping the buck during this time, there is still a heavier loss to be considered —that of the does which have already been mated, and may not prove to be in kindle, or which have produced, as a result of the bad condition of the buck, litters which are small in size and number and slow-growing.

Feed the male carefully, then, give him a hutch that allows him to exercise his limbs, and if you can let him have a run about the shed floor, backyard, or garden for an hour or two occasionally so much the better.

The moral of all this is : Before you hastily condemn your does for proving sterile or producing small and weak litters, make sure of your buck. Watch carefully the results, and if they are uniformly unsatisfactory either buy another male or borrow a buck whose reputation is good. If the matings are then far more productive give short shrift to your own buck—unless you have been neglecting him in any of the ways already mentioned.

MATING

The first rule in mating rabbits is to run an elderly buck with a young doe or *vice versa*, not two youngsters together. Rabbits *may* be bred from at four months old, but this is very bad policy. They should be at least six months old, better seven months, and better still eight or ten months old, before they are allowed to take on family cares, and never over two and a half or three years old. A good mating would be a doe eight months old and a buck eighteen months old.

A doe will only receive the attentions of the buck when she is in " season." The signs of " season " are easy to distinguish—restlessness, stamping or thumping of the hind feet heavily upon the floor, scratching at the corners of the hutch, plucking fur from the chest, carrying about mouthfuls of hay, and frequently a show of temper. When these symptoms are observed she should straightway be placed in the buck's hutch, *not* the buck in the doe's hutch, as is frequently the case with beginners, and results watched.

Directly the buck has finished his attentions (which, from first to last, will have only occupied a minute or two) he will fall over upon his side, and then the doe should be removed. Some breeders, in their anxiety to ensure a

successful mating, allow two, three, and sometimes four
" falls," but the following table, based on experiment,
proves that one fall is sufficient :

No. of falls.						No. of Young in Resulting Litter.
1	..	..	..	..	..	.. 10
1	..	..	..	..	..	.. 17
2	..	..	..	..	..	.. 5
1	..	..	..	..	..	.. 11
2	..	..	..	..	..	.. 12
2	..	..	..	..	..	.. 12
1	..	..	..	..	..	.. 13
2	..	..	..	..	..	.. 12
1	..	..	..	..	..	.. 7
2	..	..	..	..	..	.. 8
1	..	..	..	..	..	.. 10
2	..	..	..	..	..	.. 9
2	..	..	..	..	..	.. 12
2	..	..	..	..	..	.. 13
4	..	..	..	..	..	.. 1
5	..	..	..	..	..	.. 3
2	..	..	..	..	..	.. 7
4	..	..	..	..	..	.. 2
3	..	..	..	..	..	.. 4

TO HURRY ON BACKWARD DOES

If any doe is backward in coming into " season " she
may be quickened in two ways. First try placing her in a
vacant hutch previously occupied by a buck. Many stub-
born cases of backwardness have been cured in this way,
but if it does not have the desired effect the doe may be
given maple peas soaked for forty-eight hours in warm
water until they have commenced to sprout.

But such stimulating food must be used with care, as it
causes overheating of the blood and consequent skin erup-
tions. I know in some cases that sprouted peas have been
given in large quantities for quite long periods. This is
both unnecessary and undesirable. A handful on each of
three or four mornings will invariably be successful.

When a doe, after showing signs of season' and being
introduced to the buck, refuses to accept his attentions
within a reasonable time she should be removed and tried
again the following day. They should on no account
be left together all night, as there is a great risk of their
savaging each other.

There is one other point to be dealt with before I leave the question of mating—the length of time which should be allowed to elapse after a doe has littered before she is again put to the buck.

Some breeders mate a doe that has cast a dead litter two or three days afterwards, but this is very cruel. I believe in allowing at least two weeks. Given this time, they have every opportunity to fully regain their strength.

THE PERIOD OF GESTATION

The period of gestation with rabbits is between thirty and thirty-one days, though with an aged doe the time may be extended to thirty-two or in exceptional cases thirty-three days. The best way to arrive at the date on which the litter may be expected is to ignore the day the rabbit is mated. Then thirty-one days from that date will find the litter arrived. Occasionally maiden does will be one day later.

During the whole of the gestation period the doe should not be disturbed unnecessarily or in any way frightened, and all precautions should be taken to exclude mice or rats from the hutch in which the doe is housed. Worrying by strange dogs or cats is a frequent cause of litters being premature, and at times forsaken.

Three weeks after being mated the doe will be seen to be plucking or stripping the fur from her chest, flanks, and belly. With this fur she will line and cover the nest. At the same time the stripping of the fur from her belly has the effect of baring her teats, and thus the young have easy access to them. To supplement the fur plenty of hay should also be provided for the nest. The doe will carry the hay herself to where she is building the nest. This carrying about of fur and hay makes her very, very thirsty, and unless provided with something to drink she will probably devour her young when they arrive. Lack of water, indeed, is the principal reason for rabbits destroying their young.

The water should be renewed once daily at least. If a little milk can be mixed with it so much the better.

Two or three days before the litter is expected it will be necessary to clean out the hutch thoroughly, for this will be the last occasion on which you have a chance to do the work for a considerable time.

Remove the doe from the hutch very carefully. With one hand firmly grasp the ears, and let the other hand carry the weight of the rabbit by placing it (the hand) under the haunches.

WHEN NEST BOXES ARE NECESSARY

If there is no dark compartment to the hutch it will be necessary to provide a nest box soon after the doe is mated.

This box should not be less than 16 inches long, 12 inches wide, and 8 inches deep. A top to the box is optional. If the doe is an extra large one the box may, with advantage, be a little bit larger, your guide in this matter being to provide one large enough for the doe to get in and out of with comfort.

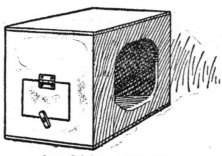

A useful form of Nest Box.

Place the box in the darkest corner of the hutch, for that is the place most favoured by the rabbits.

The above sketch shows a useful form of nest box. The little door at the side is not absolutely necessary, but it enables you to have a peep at the litter without disturbing the mother.

THE ARRIVAL OF THE LITTER

On the day the litter is due, leave the doe severely alone, but double your precautions to protect her against disturbance by cats, dogs, and other worries. If you feel you really must examine the litter you may do so at the end of the first day, but it is far safer and better to wait two or three days. If there are any dead youngsters before that the mother will invariably deposit them in the outer part of the hutch.

Don't expect anything very startling when you take

your first peep at the nest. All you will see is a moving mass of fur, but in a day or two you may note a wonderfully fat little bull-head, with fast-shut eyes (rabbits are blind for the first fortnight), poked through the fur to enjoy a snack of fresh air.

Before making a closer examination of the nest, coax the doe away from it with some little tit-bit that will occupy her attention for two or three minutes. Alternatively you may remove the doe from the hutch altogether while you make your examination, the step you take depending on her temper. Also it is a wise precaution to rub your hands with a little of the damp sawdust in the hutch. A rabbit's scent is very keen, and if her nose tells her that you have been handling her babies she may do them harm. The sawdust will take away the " human " smell from the hands.

When all is ready, open the nest quickly and cleanly and at once remove any dead youngsters. Then count the remainder, and if there are too many for the doe to rear, either kill the surplus—of course choosing the smallest and weakest-looking for destruction—or foster them on to another doe, as described later on in this chapter. From four to seven young ones may be left, according to the size and strength of the mother.

Your examination completed, rearrange the nest as nearly as possible exactly as you found it, return the doe to the hutch and give her another tit-bit to soothe her. The mother and her family should then go on nicely with little more attention from you, save an occasional glance to make sure that no more have died, until weaning time. The doe knows better than you do what is good for her young, and when the time is ripe for them to make their appearance before the world she will stage-manage that event. Sometimes they will not leave the nest for three weeks, but usually you can see them hopping about at the end of a fortnight.

The litter makes a great demand upon a doe and everything should be done to encourage a good flow of milk. When a litter is born dead or dies soon after birth, however, non-milk-producing foods should, of course, be resorted to. Middlings in which half a teaspoonful of salt has been mixed is as good as anything, while green food should be entirely withheld for several days.

On rare occasions, generally during the summer months, young doe rabbits will show a desire to mate when their litters are very young, and as a consequence cause annoyance by harassing the young. When such is apparent the doe should be removed and may be mated forthwith, after which procedure, if she is kept quiet, she will often behave herself when replaced in her original hutch. If, however, she appears spiteful, keep her away altogether. Such instances are rare, but worthy of record as a passing hint.

WEANING THE LITTER

The youngsters will be noticed nibbling at the mother's food almost directly they have left the nest, and this fact gives many breeders the opinion that they are then able to shift for themselves. Nothing of the sort. A very large number of rabbits have been lost as a result of premature interference, while others mature very slowly.

This slowness of growth is not entirely attributable to their withdrawal from the supply of milk they obtain from the suckling doe. The warmth, particularly at night time, that they receive from their mother's body combined with other attentions, such as cleaning their fur, etc., all assist the litter towards a quick growth, and early development of size, two very essential points worthy of attention.

Six weeks in summer and seven in winter is about the right age at which they may safely be left to shift for themselves, though occasionally it may be necessary to remove the doe at an earlier period as, for instance, when the doe has been brought to a poor state of health through the excessive strain of suckling a big litter. Usually, however, when this happens relief may be given merely by taking half the litter away and encouraging the youngsters that are left to feed themselves as much as possible. A pan of milk or soaked bread and milk will be found excellent for this purpose, and may be given with advantage twice daily. The feeding will not only help them in their growth, but, at the same time, will greatly assist in relieving the doe.

If in a few days the doe does not improve in her condition, however, then the remaining youngsters must also be taken from her.

FEEDING EARLY-WEANED LITTERS

The litters that are weaned thus early will need very careful feeding, or the sudden change from nest feeding to a diet of ordinary foodstuff will cause digestive troubles leading to fits, scours, etc.

For the first few days only give substances that are easily digested, such as warm bread and milk, sweet, soft meadow hay, and a little tender green food, all coarse or rank green-stuff being avoided. The amount to be given should be just enough to enable them to clear it up without leaving any over. Meadow hay is the exception, and may always be before them to nibble.

The secret in managing these litters is to feed often and regularly. I have obtained excellent results by arranging the hours of feeding during the first week of weaning at the following times, 6 a.m., 10 a.m., 2 p.m., 6 p.m., and 10 p.m. After the first week the feeding may gradually be reduced to three times daily, and the amount of food increased according to their appetites.

JUST BEFORE WEANING

Even with litters that are to be weaned normally it will pay you to give a little extra attention to the feeding, giving the youngsters the kind of food it is intended they should have after they are weaned, though naturally preference should be given to bits that are tender and easily digested. Not only does this accustom the litter gradually to the coming change, but it relieves the doe, whose milk will naturally begin to fail after the first few weeks.

To allow the litter to suffer on account of this diminution of their natural food would be most shortsighted policy, for it is at this period that the framework of their bodies is built up, and on this will depend to a great extent quick development. A check in growth may mean that before they are of a marketable size it would be necessary to keep them quite a month longer than would otherwise have been the case.

When the youngsters are fully ready to be weaned it is far better to remove the mother from the hutch than to remove the litter, if the hutch is fairly roomy. A change of quarters often causes a check to growth.

On no account should the litter be placed in separate hutches. It is a big change for a youngster to leave a litter of six or seven others, and if it is necessary to change the hutch it is far preferable to let several of them run together for a while in their new quarters. All nervousness and inclination to refuse food is thus forgotten, and by the removal of these two objections the rabbits continue to grow.

THE GROWING STOCK

The aim of the table-rabbit breeder should be to have his stock at the right killing weight when three months or at least sixteen weeks old. It must be remembered that the longer they are in hand the more food they consume and the less profit they show. Careful, regular, and plentiful feeding should, therefore, be the rule.

Given this, the youngsters will grow very fast, often putting on over half a pound a week. I have actually had four months' old rabbits weighing 6 lbs. (live weight). The truth is, rabbit fattening really begins with the doe's milk, and if their appetites are kept keen by frequent changes of food there is often no need for any special fattening diet.

For the last few weeks of their lives, when they have put on considerable weight, it may be found that the youngsters' quarters are somewhat cramped, and cramped quarters are all against quick growth. A change will therefore be necessary. If a large hutch is available this will answer the purpose very well; if not they may have the run of a shed provided it is well ventilated, but not draughty or damp. Too much freedom, however, and consequently a large amount of exercise, may restrict the fattening, and if the shed is a large one it would be as well to pen off a section by means of wire-netting.

When several litters are run together in this way, considerable labour is saved, and this is a consideration when fifty or a hundred rabbits have to be attended to.

Feeding will be a little difficult at first, because the stock will vary greatly in the amount of food they consume, but observation will quickly teach you the right quantity.

Before killing off a litter always look them carefully over on the chance that there is one worth keeping for sale or stock. A Flemish Giant, for instance, that is a good steel-grey all over, including the feet, is always worth keeping, because such are not bred every day. Those Flemish, too, which have white' bellies will also turn out better than those showing no white on any part of their bodies.

It is always advisable to keep the pick of the bucks from an early litter either for use in your own stud or, if new blood is wanted, for exchange against a buck equally as good belonging to a neighbour.

The introduction of new blood now and again is always good policy. Of course, when raising utility stock you must on no account inbreed. The buck must not be even distantly related to the does he is run with.

Young rabbits that are being kept on for stock or fur must be separated as to sex when they are fourteen or sixteen weeks old, according to the breed and the time of the year. Bucks must be run with bucks and does with does.

FOSTER-MOTHERS AND FOSTERING

I made a brief reference to the rearing of surplus young by means of foster-mothers. It is surprising to what little use foster-mothers are put by rabbit breeders. Some have never heard of fostering, others look upon it as purely a trick of the " fancy." But while of inestimable value to the rearer of fancy stock, it is also invaluable to the utility breeder, and should be practised by all whenever occasion demands, as it enables considerably more young-sters to be reared to marketing age than would otherwise be the case.

The first step to take is to mate several does on succes-sive days, so that they will all kindle round about the same time. When the litters arrive you examine the nest as before directed, and find the number in each. Some does may have nine or ten, others only three or four, and the principle is to divide up the youngsters so that each mother may have an equal amount of nursing to do.

Coax the mothers away from the nests with some favoured tit-bits. Then take from the large litters the number of young over and above five or six ; next take up one of the youngsters from a small litter, and holding it by the fore part of the body, gently squeeze it, so making it pass urine. Allow the urine to drop on to the youngsters which you intend to introduce into the same nest as it is occupying, and gently rub them all together so that they will all be smeared over with the urine. Do this with each batch of young to be fostered.

Rabbits recognise their young by the smell. The procedure described gives the new-comers the same scent as the original occupants of the nest and the mother will not notice the increase in her family.

When raising utility rabbits on a large scale it is an excellent plan to keep one or more does that are known good mothers purely and simply for fostering, mating them on the same day or thereabouts as the remainder of the stud, but destroying the young off-hand, if the size of the other litters demand it, or as many as occasion requires.

The Dutch variety takes pride of place as the best foster-mother. Despite the fact that they are small, they show a wonderful devotion to their offspring, and it is only on very rare occasions that a Dutch doe will fail in this part of her duty if given only ordinary attention. She will also usually allow almost any interference with her nest without alarm.

The does of the English, too, are very trustworthy, and an experiment at fostering can be safely undertaken with this breed if no Dutch are kept.

It must not be inferred that no other breed—or cross breeds—answer the purpose. The two varieties named have been mentioned solely because they are preferable for the purpose. A point worthy of consideration is also the probability of breeding a first-class specimen from the Dutch or English if they have been mated to a pure-bred buck.

WINTER BREEDING

For the reason that spring is the natural breeding time for *wild* rabbits, many people jump to the conclusion that to breed *utility* rabbits at any other time is extremely difficult. This is a mistake. Utility rabbits will breed all the year round, and will rear their litters successfully provided proper precautions are taken during the colder months of the year.

1. FEED THE DOE ON CONCENTRATED FOOD & SHE WILL MATE READILY.

2. A SACK HUNG OVER FRONT OF HUTCH HELPS TO KEEP THE COLD OUT.

3. IN BITTER WEATHER A HOT WATER BOTTLE - CAN BE PLACED NEAR THE NEST.

4. YOUNGSTERS CAN BE LEFT A GOOD TIME WITH THE DOE.

Some of the little extra attentions that the wise rabbit-keeper will always give his rabbits when they are breeding during winter-time.

These precautions include—

1. The provision of a thoroughly sound and weather-proof hutch, which can be additionally protected by means of a screen during heavy rains and cold winds.

2. Good stock, perfectly fit in health.

3. The provision of ample and warm bedding material and a plentiful supply of food.

4. The reduction of large litters so that the mother may give sufficient care and attention to all, particularly at night.

In cold weather the hutches may be artificially warmed by means of a hot water-bottle or hot brick, wrapped in

E

flannel and placed flush up against the nesting compartment, outside.

Half-gallon jars of the style and shape used by herb beer hawkers are just the thing for this work. If filled with boiling water and wrapped round with some old material, it is astonishing what heat they give off and the length of time they retain it.

It is often possible to get them almost for nothing at hotels or large public-houses, where they get the handles broken off them and are then put aside. It is hardly necessary to add the desirability of using a good tight-fitting cork.

Bricks heated in the oven and used in the same way serve a similar good purpose.

Personally, however, I only use artificial heat when I notice a doe anxiously carrying about her offspring from place to place in an attempt to find a cosy spot. The average utility rabbit is a hardy specimen and can put up with any amount of cold, though draughts and damps may work havoc. And I don't believe in coddling any live stock.

Apropos of having the stock perfectly fit for winter breeding, it is not advisable for *maiden* does to be mated after the end of October, as quite 50% turn out to be indifferent mothers when their first litters are dropped in cold weather. Therefore, to obtain the best results from winter breeding, only those does that have previously had a litter should be used.

The most suitable bedding is, first, a thick layer of sawdust placed over the floor of the hutch and then an ample supply of straw.

So far as feeding is concerned, it is only necessary to remember that the stock naturally have keener appetites in winter than in summer. The ordinary feeding may be practised up to a certain point, but the last thing at night it is advisable to give them an additional good feed of haystuffs or hay substitutes.

Rabbits that are housed in warm, comfortable sheds consume less food, and at the same time will be found to grow much more quickly, table rabbits being ready for market several weeks earlier than those that are housed in outside hutches. I have also frequently noticed that during spells of foggy weather pneumonia is more prevalent where the stock are kept in the open. The moral is obvious. If possible bring your hutches into a shed or outhouse during the very severest weather.

Remember that the buck requires every bit as much attention as the does. He must be equally well fed and kept in good condition generally, and on no account be used more than three times a' week. If the buck is made use of more than three times a week, there is a possibility that he will lose his vitality. Loss of vitality is largely the cause of so many of those weakly litters which we see each year as the result of winter breeding.

CAN THE SEX BE INFLUENCED?

The subject of sex influence is one that has from times far back agitated the minds of breeders and led to statements, said to be reliable, but which have nevertheless again and again proved to be purely speculative suppositions. What I have to say is not based on theoretical principles, but is the result of actual and consistent application of a system adopted by a well-known breeder purposely to see if Nature could be cheated by breeding male or female progeny at will.

A very exact and accurate record of every rabbit bred was kept, and therefore he was in a position to trace the ancestry of any particular specimen referred to.

The breeder does not claim to have found an infallible system which once and for all settles the question of sex influence, but he does claim that in the majority of instances when he has applied his principles he has obtained more or less specific results.

From a mating taken *directly* the signs of season are apparent, he found on examination of the resulting progeny that the male sex predominated.

For three successive matings he carried this out carefully, and the results obtained were as follows:

Mating No. 1. Out of five young there were four bucks and one doe.

Mating No. 2. Out of seven young there were five bucks and two does.

Mating No. 3. Out of six young there were five bucks and one doe.

The three matings provided results so similar that I consider they deserve publicity. The first and last of these matings were by the same buck, whilst the second was by one of similar age purposely introduced.

At a later period he experimented again, this time about four days after he first noticed the doe desired to breed. The same bucks and in the same order were used as previously.

Three matings this time brought the following results :

No. 1 mating, six does, two bucks.

No. 2 mating, seven does, two bucks.

No. 3 mating, five does, three bucks.

Although not quite so successful as the first three matings, the result seems to point in a sufficiently marked degree that the sex problem can—at least in a measure—be regulated, and as there is always a greater demand among breeders for does than bucks, I consider the experiments of great importance.

CHAPTER V

GENERAL MANAGEMENT

RABBITS are less trouble to look after than almost any other stock. Provided all goes well, the only regular labour involved is the feeding of the stock and cleansing of the hutches. Isn't it surprising, then, that because so many people don't take the trouble to clean out their hutches regularly rabbits should have got the reputation of being "smelly" animals ?

Rabbits are about the cleanest live stock of all and the least likely to smell. They are about the only animal which will use one place alone for sanitary purposes, and will keep the remainder of their living quarters dry and clean. Naturally if you leave the droppings to accumulate and the litter to become a soaked mass there is bound to be an unpleasant odour ; but if you take the trouble to clean out the "favoured corner" every day and the rest of the hutch twice a week, you couldn't tell there was a rabbit within a mile of you if you had only your nose to guide you.

THE VERY BEST LITTER

I am a very strong believer in sawdust for use in the rabbitry ; I have found it to be the very best litter obtainable. It is a disinfectant, a deodoriser, and a deterrent to insect pests and soaks up all the urine. Not only are the floors of hutches also preserved by its use, but cleaning-out is greatly facilitated. With the aid of a scraper and shovel, the refuse of a dozen hutches can very quickly be removed, leaving them sweet and clean for further occupation. It is cheap—you can get a large sackful for a shilling if you live near a sawmill—and you can therefore be lavish in its use. But it will be necessary to cover the whole of the hutch floor with it. Put a layer an inch thick in the sleeping compartment and a good handful in the corner frequented by the rabbit and that will be sufficient.

When sawdust is used, the hay or hay substitutes—straw, bean and pea haulm, etc.—which should always be given to provide warmth, will be treated just as much as food as bedding, and will therefore serve the dual purpose.

In country districts, where sawdust is quite unobtainable, earth that has been thoroughly dried in the sun or in an oven and stored in a dry place is a good substitute.

I like to mix a little disinfectant with both sawdust and earth, and this is certainly a cheap insurance against disease.

In many country districts supplies of bracken are easily obtained in the winter-time, and this can with advantage be used on the top of sawdust. In such cases, however, it is advisable to give a handful of hay per rabbit each day. They always relish it, and stocks that get supplies of hay always carry plenty of firm flesh.

Wood shavings are also useful as a bedding, but here also hay or hay substitutes must be given for food.

CLEANING OUT THE HUTCH

As I said before, it is necessary to clean out the whole hutch twice a week and the corner that is most used every day. To facilitate cleansing there is an excellent little tool which every rabbit-keeper should possess. It is a small triangular iron scraper, and costs from threepence to sixpence, if not made at home from a stout piece of tin or a bit of sheet-iron.

A small stiff-bristled brush for sweeping out the hutches is another useful accessory, while a dustpan will prevent the floor around the hutches from becoming soiled and littered with rubbish. Sweep all refuse into the pan and throw it into a heap. It will be very valuable as a garden manure.

The right way to clean out a hutch is to shake the hay or litter to allow the dung to fall through, remove hay that is soiled by urine and stack it in a corner. Then rake any droppings from the dry sawdust and sweep the latter into another corner. Next use your scraper on the most frequented corner and scrape out all damp sawdust, droppings, etc. Throw a heap of sawdust into the soiled corner, replace the sawdust on the floor of the sleeping compartment, rearrange the bedding, and the job is finished. Of course, you must make good any litter that has been eaten by the stock.

When the does have young ones under three weeks old it is as well to leave the cleaning of the breeding compartment either alone altogether or until the mother is very closely occupied in feeding.

THE FEEDING ARRANGEMENTS

It helps to keep hutches clean and tidy if all food is fed to the rabbits in troughs. In the matter of feeding troughs we are now much better off than we were a few years ago when the old-fashioned wooden kind were practically the only sort in use. Besides being impossible to keep clean and therefore liable to cause injury to health there were all sorts of other bothers associated with these troughs. For instance, the rabbits invariably regarded them as a means for keeping their teeth in order and in a short time had gnawed them to pieces. Also, if they were not fastened down they were often to be found on the top of the nest containing the youngsters.

Up-to-date rabbit-keepers now insist on having the earthenware kind of feeding trough with the flanged edge, for these can be scalded when necessary, and the flange prevents waste, whilst their weight prevents the rabbits from turning them over. Two types of these are illustrated on page 33.

A few years ago I had several troughs specially moulded by a brickmaker. They are the size of, and exactly like, a brick except that the centre is, of course, hollowed out. I find them very excellent things for a hutch containing a litter of growing youngsters. They cost me 6d. each. The man who made them is now, however, unfortunately (for me) out of business or he could have had further orders. Any reader, however, living in the vicinity of a brickworks will most probably be able to get one or two similar troughs made.

Metal troughs can also be bought, and those types which are attached to the wire of the cage are convenient. Their only disadvantage is that they work out rather expensive.

One breeder I know uses the tins in which 1 lb. of gold flake tobacco is packed, measuring 7½ inches by 4 inches by 1½ inches. He removes the lid, punches a hole midway in the crease near the top, and with a short screw fastens it, not too tightly, to the inside of the door of the run, so that it just clears the floor when the door is closed. It is then very handy, for if, when feeding, there is anything to clear out, the door being opened, a half-turn will drop the contents into anything placed underneath.

If you must use wooden troughs, at least screw two strips of tin along the tops of the two sides, as shown in the sketch on page 33. This will stop gnawing.

For water the earthenware food trough, already referred to, an earthenware garden seed pan or a tobacco tin, fixed as described above, will all answer the purpose. Let me once again remind rabbit-keepers that all rabbits, not only breeding does, *do* require drinking water and should have a fresh, clean supply regularly before them, no matter what you have heard to the contrary.

If you doubt that rabbits drink, put a cup of water into one of your hutches and note the actions of the inmates. That will convince you.

THE FOOD STORE

One of the elements of success in profitable rabbit-keeping is to buy foods in bulk. When this is done, however, it is necessary to provide some means of storing the foods. Whether or not a corn-bin is used can depend on the particular circumstances which obtain in any one rabbitry, such, for instance, as room to accommodate it. Some contrivance, however, must be brought into use that is mouse-proof, for if this is not so, apart from the oats, etc., getting soiled, a great amount will soon be consumed if there are any of these rodents about.

Many breeders use large tin boxes with tight-fitting lids

for both oats, bran, sharps, etc., but I use small zinc bins. The outlay and upkeep of what might be termed the " rolling stock " of rabbit-keeping is far less than that necessary for any undertaking in connection with the keeping and rearing of other kinds of live stock, and there should be no reluctance, therefore, on any one's part in providing suitable and sensible specimens of those items that are essential both for the comfort of the stock and the convenience of the owner.

MANAGEMENT DURING THE MOULT

The moulting period is a critical time with rabbits, and each year large numbers are lost through either neglect or carelessness. During the moult the rabbits, of course, change their coat, and though some animals do so more easily than others, on all during this time and until the new coat is fairly established, there is a heavy strain, and unless they are well looked after the breeder is sure to suffer a loss sooner or later.

A specimen may get half-way through the process and then seem suddenly to stop. This is spoken of by breeders as " sticking in the coat." When a rabbit has difficulty in shedding its old fur, it can be greatly assisted by a daily grooming with the bare hands.

The chief ailment that affects rabbits during the moulting period is a cold. It is generally brought on either by a sudden change in the weather or by a draughty hutch or rabbitry. Although it is necessary to exclude draughts, do not overlook the fact that to prevent colds it is most essential that ample ventilation should be provided.

The most important thing to consider after ventilation is the feeding. This should consist of the very best materials, and must be given in sufficient quantities.

It must be understood that in addition to the ordinary functions of the body, the animal has to supply the material for growing the new fur, etc., and generous feeding is required to supply these wants. Unless it is forthcoming the rabbit will lose strength and then will be liable to take cold very quickly.

Too great care cannot be exercised in preventing colds.

It is a cold in the head that may develop into that most dangerous and highly infectious disease, snuffles, which is the worst of all rabbit diseases.

In addition to the ordinary foods fed to the stock, maple peas will be found excellent as a pick-me-up. They should first be soaked for about forty-eight hours and then allowed to sprout. A small handful given three or four times a week will be found sufficient. Linseed cake is another good food. It should first be broken up very small and to prevent waste should be fed in proper troughs or feeding pots. Linseed meal is another very useful article for assisting the coat into condition. As it has the tendency to loosen the bowels, however, it should be used sparingly.

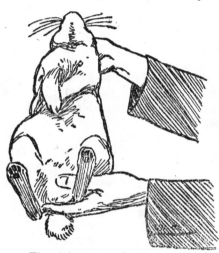

The right way to hold a rabbit.

On no account is it advisable to allow the stock to mate when in moult, as the offspring from such a mating will generally be found to be poor and delicate.

HOW TO PICK UP A RABBIT

Different breeders have different views on the way to pick up a rabbit, but one and all are now agreed how *not* to do it—solely by the ears. This is not only extremely painful to the animal and therefore cruel, but it spoils the set of the ears.

The right way is to take hold of the ears or the scruff of the neck with one hand, and place the other beneath the body close up against the hind legs, as shown. You can carry it about conveniently in this way, and the rabbit suffers no pain.

If you want to hold a rabbit very firmly so that it can be examined, take hold of the ears close up against the skull firmly, the thumb in front, the fingers behind (*see illustration*). It is then quite powerless, and though a hand may be placed against the back to support it, no grip is necessary with that hand.

THE BUSINESS SIDE OF RABBIT-KEEPING

The old fallacy that rabbit-keeping can only be regarded as a pleasant hobby has long been exploded ; yet there are still many breeders who are not in the position to say definitely what their profits or losses are on the past season's working. But it is only by proceeding on strict business

If you grip the ears of a rabbit in this way it will keep perfectly steady.

lines and adhering to strict business principles that rabbits can be made to pay handsomely.

Let your rabbit-keeping become a business with you, no matter on how small a scale it is conducted. Keep an account of everything that is spent or received for rabbit transactions.

A penny notebook showing the dates of matings, the number of youngsters reared by each doe, how they have been disposed of, and the profit made, makes very interesting reading at the end of the season, but if no such record is kept, what guide has any breeder as to what his future policy should be ?

CHAPTER VI

WOOL RABBIT-KEEPING

THERE is a branch of utility rabbit-keeping which has not so far been touched upon in this book, but which none the less deserves the consideration of all—the keeping of wool-bearing rabbits. Rabbit's wool is now an extremely valuable product, being used for various purposes, the demand for it, therefore, always being steady, often, in fact, exceeding the supply. The price, of course, fluctuates, but at the time of writing it is round about 40s. a pound.

As the average yield of wool from each rabbit is 12 oz. a year, it will readily be seen that wool rabbits are a profitable proposition.

The one and only breed to keep for wool purposes is the White Angora. This is a beautiful rabbit, with a pelt so fine and silky that it may aptly be termed wool. Often the depth of the pelt is 9 or 10 in., snow white throughout. In show trim, an Angora resembles nothing so much as a monster powder-puff.

There are other colours than white—blue, smoke, and fawn—but they belong more to the fancy than the utility side.

Incidentally the pelt of the White Angora, as a pelt, is also saleable at a good price, and as the flesh, too, is good enough to justify the inclusion of the breed as a table rabbit, you have a triple combination which cannot be equalled by any other rabbit. The ordinary fur rabbits have a dual purpose only.

Lest you come to any hasty decision regarding the breed you should start with, however, let me say that the Angora is considerably more trouble to the rabbit-keeper than either double-purpose or single-purpose breeds. It *must* have daily attentions which are not needed by any other rabbit, and cleanliness is more than ever important. Angora fleece, you see, is so readily stained, and stains are so difficult to remove. Stained fleece, of course, brings in only a poor return.

For this very reason it is best to keep Angoras in an indoor rabbitry or, at least, to have a permanent awning over the hutches. When rain drives into the hutches, or when the hutch interior is exposed to damp fog and mists, it is practically impossible to have the litter sufficiently dry and sweet to keep the wool in that state of perfect cleanliness that is so necessary.

In order that the fleece may not be damaged, it is advisable to have the hutches even more roomy than usual and to pay particular care that there are no nails projecting from the woodwork or no sharp ends of wire from the netting. I prefer to use wire rods in preference to netting for the hutch front, in fact.

Coming to the routine work with this breed, of first importance, then, is the *daily* cleaning of the hutches and the *daily* renewal of the litter. As the wool is liable to pick up litter and thus become matted, the best scheme is to keep a thick layer of dry deal sawdust on the floor, covered with a thin layer of clean oaten straw cut into 8-inch lengths.

The management item of most importance is daily grooming. If you once let the fleece get out of hand it will become so tangled and matted as to be practically valueless. That is one reason why Angora keeping is not the side-line for those of careless or slipshod habits. Without the slightest doubt such would find the stock a dead loss.

Here is the proper way to groom :

Take a long-haired soft brush of convenient size and gently brush down the rabbit, starting at the head, working to the tail, then brushing towards the head and outwards. If any tangles are met, don't rest content until they have been unravelled, and don't call in the aid of a comb unless you must. A certain amount of wool is bound to come away at each brushing ; this can be saved, being kept dry and clean until a sufficient quantity has been saved to justify its despatch to a buyer.

Grooming is best done whilst the rabbit rests on a bench to which a piece of carpet has been tacked.

As to feeding, this may be on the ordinary lines, but naturally the food supply should be generous. Good feeding promotes quick growth and is conducive to large size, obvious advantages with wool rabbits as with all other classes. Angoras, for some reason or other, suffer more from bowel disorders than other breeds. Such should, therefore, be guarded against—by feeding plenty of hay, avoiding stale or poor greenstuff, and by never changing the food abruptly.

When starting an Angora rabbitry it is most important to buy good stock. Far better to spend the money you have to spend on three good animals in preference to half a dozen of inferior quality. If adults are purchased they should be of good size, bodily, and should possess a good depth of fleece. They should weigh at least 6 lbs.

The does are invariably good mothers and are fairly prolific, averaging anything up to eight youngsters in a litter. They will breed freely all the year round without undue trouble, provided attention is paid to the special winter breeding points enumerated in another chapter.

In the matter of breeding, however, it is *particularly* important to see that both the buck and the doe are in good coat at the time of mating. If either of the parents show the slightest trace of moulting at the time, the youngsters will never be really well fleeced ; their coats will always be coming out ; no amount of care in grooming will keep them trim.

There are two methods of obtaining the wool " harvest " —by plucking and clipping. Clipping is the commoner method in this country, though in France, where Angoras are kept on an enormous scale, almost on rabbit-farm lines, plucking is usually practised. Provided the plucking is done when the wool is ripe, it is quite painless, the down coming away almost without effort on the plucker's part.

Clipping is usually done three times a year, spring, summer, and autumn. Plucking can be done at any time when the wool parts easily. The usual procedure is to semi-pluck each rabbit about every six weeks, taking the wool from the back of a specimen only at one operation, and then, at the next, when the back wool is half grown again, plucking from the under part of the animal.

Plucking can commence on youngsters when they are

three months old, but should not be repeated for three
months, another three months' interval elapsing before
the third plucking takes place. If the skin reddens under
the plucking operation—in adults as well as youngsters—
the wool is not ripe and pain is being caused the animal,
which should accordingly be left a while longer.

It is never advisable to pluck rabbits completely except
during the height of the summer, this for the obvious
reason that they would be liable to catch cold were you
to do so. Neither is it advisable to pluck does later than a

Some of the most important points to note in connection with rabbits' wool production.

month before they are due to kindle or before the resulting
litter has been weaned.

The wool *can* be marketed in half-pound lots,[1] but it is
better to make up packages of at least a pound in weight.
The greatest care should be taken of the wool whilst the
lots are being accumulated. If allowed to get damp or
become soiled its value will rapidly deteriorate. There
are first, second, and third qualities in Angora wool, and
naturally only the first quality commands top price.

[1] If you wish to know a buyer write the Editor of *The Smallholder*,
18 Henrietta Street, Covent Garden, London, W.C., who will be pleased
to supply names and addresses.

CHAPTER VII

RABBIT COURTS AND RABBIT WARRENS

I HAVE frequently had such questions asked me as: " How many rabbits must I keep to bring me in a profit of £1 a week ? " the questioner having in view a rabbit farm run on very similar lines to a poultry farm. My invariable answer has been : " You couldn't do it," for I have never favoured the rabbit-farming idea. When you have to buy or grow large quantities of rabbit food, rent land, employ labour, and keep a big stud of rabbits in tip-top breeding condition, all sorts of outside questions arise that don't crop up at all when rabbits are treated as a side-line and fed for next to nothing on garden waste and wild food.

But if rabbit-farming is a risky speculation there is a step above ordinary hutch rabbit-keeping which is usually very profitable—rabbit courts and rabbit warrens. Such, of course, are only suitable for rabbits destined for the table. Fur rabbits are best kept under the ordinary system.

A rabbit court and a rabbit warren must not be confused, for whereas in the latter case the rabbits are allowed to burrow and have their freedom, the court only allows them restricted liberty in a closed-in yard, giving them hutches to sleep in.

The yard should be as large as possible, and paved with bricks, large stones, or other material so that the rabbits will be prevented from burrowing. A boundary on each side will be necessary, sufficiently high to prevent the rabbits from jumping over (at least 5 feet). If a place exists—as there does in numerous rural districts—where a good wall bounds a yard, so much the better, as cold winds are kept at bay during the winter-time.

If no walled-in enclosure exists, a wire-netting fence,

The White Angora, the one and only
breed to be kept for wool production.

To face page 81.

5 or 6 feet high, might be erected; but the situation chosen should be a warm one, preferably facing south. A quick-growing hedge might also be planted on the exposed sides to provide additional shelter.

The does' hutches should run along one side at the foot of the most sheltered wall, and the bucks' hutches as far away from them as possible. The does' hutch may be the ordinary two-compartment breeding hutch, but the buck will only use his for sleeping and therefore no exercise compartment will be necessary. These hutches must be cleaned out regularly once a week and fresh bedding material put down. It must not be thought that because the rabbits have their liberty the hutches will remain clean indefinitely.

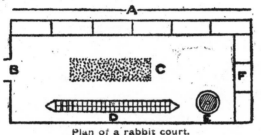

Plan of a rabbit court.

A. Does' hutches.   B. Entrance to court.   C. Mound of sand for burrowing.
D. Feeding trough.   E. Drinking fountain.   F. Stud bucks' hutches.

Some kind of occupation must be found for the rabbits in the daytime, and a load or two of sand or other similar material tipped in the middle of the court will be undoubtedly very much appreciated. The rabbits find untold amusement by burrowing and frolicking about in it. Care must be taken, however, to see that they are not allowed to sleep out in these burrows in the winter, or they may be frozen to death.

The accompanying plan shows a well-arrranged court.

There are all sorts of breeds that might be turned down into the court—the Silver Grey, the English, the Lop, etc.—but I still remain on firm ground when I say that there is nothing to beat the Flemish Giant and Belgian Hare. There should be one buck to every eight does.

As adult rabbits are liable to fight, the stock intended

F

for the court should be penned up together directly they are weaned in order that they may become accustomed to one another before they have their liberty.

Feeding should take place twice or three times a day in the ordinary course of events, but during breeding times suckling does might with advantage have a little extra attention. Always when you are feeding seize the opportunity to examine individually each rabbit in the court for signs of any ailment. Directly any of the stock are noticed to be ailing they should be immediately captured and placed in an isolated hutch. If not checked the trouble might sweep through the whole of the court and leave not a single inmate alive.

RABBIT WARRENS

There is no doubt that where a suitable situation exists a specially stocked rabbit warren is a remarkably profitable undertaking. The best position is a sunny, sandy meadow, with plenty of bushes, and that there are many thousands of acres of uncultivated land of this type will not be disputed. No matter if the soil is too poor to grow food acceptable to the general run of farm cattle, it will do very nicely for rabbits, for nearly every kind of green herbage is enjoyed by them.

I will admit there are disadvantages in the system, among them expense. The principal item of expenditure will be the fencing, the cost of which will, of course, be determined by the area of ground in use. In many rural districts, however, land can be and has been acquired already fenced by hedges, etc., and this is a big consideration. A second disadvantage is that during the natural breeding season the stock cannot be controlled, for during this time they must not be interfered with. Neither can extra kinds of food be given to fatten off certain of the youngsters quickly, and this fact to a large extent is the cause of young rabbits from warrens not reaching the same plump condition so quickly as those reared in hutches.

There is also another factor beyond our control which may cause trouble, viz. the varying weather elements to which we are subjected. These at times may cause disease.

But I maintain that the advantages will outweigh the disadvantages. No constant attention is required, as the animals, of course, lead a natural life—a point very greatly in their favour. Again, a much greater quantity of rabbits can be marketed than with any other system, whilst the cost of raising them is infinitely less. A few loads of sand distributed over the land, on top of which is put rough wood twigs and faggots, will be all that is required for shelter, and the rabbits will soon make a burrow in their own way.

Again, the rabbits will resemble their wild brothers very closely so far as the flavour of their flesh is concerned, a point reckoned much of by some dealers and a disadvantage nowhere.

I am not going to say the price they realise is as much as for hutch-reared rabbits, but none the less it is far better than that received for the ordinary wild rabbit.

THE FENCING

The fencing can be put up by any man handy with tools. It should go entirely round the ground, and should be formed from good quality 1-inch mesh wire-netting. Cheap stuff is worse than useless, for in a very short time the weather will knock it to pieces. Six feet high is not too much, and the netting should also be sunk in the ground 2 feet or more. Though an additional expense, it is an advantage to use very small meshed netting below the ground-level and 2 feet above, as this keeps any prowling rats at bay.

First dig a trench 1 foot wide and 2 feet deep round the whole of the warren. At a distance of every 3½ to 4 yards drive in stout posts 5 feet high, and firmly secure the netting thereto. This done, fill in the trench and ram the soil hard. The idea of sinking the netting in this fashion is not only to make it more secure, but to stop burrowing either by the rabbits or vermin from the other side of the closed-in ground. If both netting and stakes are tarred below the ground-level they will last much longer.

STOCKING THE WARREN

From twenty-five to thirty rabbits, including six or eight bucks, is enough to put down on a warren one acre in extent. They increase and multiply at such a rate that a larger number would mean overcrowding, souring of the soil, and consequent weak and unhealthy litters. The spring is the best time to turn the stock down, but a start may be made any time during the summer.

Only young rabbits are suitable, a nice age being from four to six months. At this age there is a natural inclination to burrow.

The best breed is rather open to dispute. For some reason the Flemish Giant does not adapt itself readily to the new mode of life, but Belgian Hares have given excellent results. Silver Greys have also done well, but as the Belgian Hare always was and always will be a better table rabbit than the Silver Grey, I think preference should be given to the Belgian. You will find them very wild after a time, the unexpected liberty after many generations of captivity seeming to make them lose their heads temporarily. But if you make a point of visiting the warren frequently they will soon become accustomed to you.

This is an advantage when, during very hard weather, you have to provide additional food in the shape of a few roots.

If the warren is in a very dry situation it will be necessary to provide water. A shallow tub sunk to the ground-level (not deep enough to imprison any inquisitive youngsters who happen to tumble in) will prove a handy water store, though the water must on no account be allowed to get stale before it is replenished.

GRADING UP A WILD STRAIN

The present popularity of rabbit's flesh as an addition to the nation's larder has led a few wideawake owners of wild rabbit warrens to attempt to grade up the inmates in order that they may fetch better prices. I have been furnished with several very favourable reports of such undertakings.

In one case, a warren was nearly cleared out of its

inmates during the early spring months of 1912, those rabbits which were allowed to remain being—as far as the owner could manage—young does. To these were added ten sturdy Silver Grey bucks.

An examination the following year proved that seven of these had survived the winter and were doing well. One had been accidentally shot. It is possible the other couple may have been alive, but did not show themselves when the examination was made.

The youngsters taken from the warren were plentiful in numbers the following spring. They were generally of a greyish brown, a few being quite black all over except the under part of the tail.

It was not possible, of course, to fix their ages definitely, but they were nice specimens and carried plenty of firm white flesh under a thick coat, and sold for slightly increased prices compared with the wild variety proper.

The pure blacks were skinned before their carcases were sent to the market, but some of the grey-browns were sold with their skins on, and no question was asked by any buyer as to why they were of a slightly different type than usual.

I have no authentic report of Belgian Hares being used to grade up a warren of wild rabbits, but I should imagine the result would be even more successful. To begin with, there would be little likelihood of any blacks being thrown and then the size of the progeny should be even further improved, for the Belgian is a bigger rabbit than the Silver.

THE MORANT SYSTEM

There is one other system that deserves brief mention, the Morant system. Under this system the rabbits are housed in floorless hutches which are set on good grass. The hutches are moved to a new site every day, the rabbits eat the fresh grass thus available, and therefore live under more or less wild conditions.

The rabbits usually have a small supply of corn and an allowance of hay every day. Additional feeding is only really essential, however, during the last fortnight of a table-rabbit's life when it is being got into tip-top condition.

CHAPTER VIII

KILLING AND MARKETING TABLE RABBITS

TABLE rabbits are really in season all the year round, but the biggest demand for them is from September to April. There are three markets open to you :

1. *The large wholesale markets.* Such as Leadenhall and Smithfield in London, Sheffield, Manchester, etc. Before the war these dealt in Ostend rabbits, but nowadays deal almost exclusively in home-fed rabbits—to their great advantage, for the carcases are in better condition, having travelled only a short distance (rabbit meat travels very badly), and do not have to be disposed of so hurriedly.

2. *The local trader.* There is undoubtedly a large demand by poulterers, etc., in many big towns for home-fed rabbits, and if a fairly regular weekly supply can be promised they would be quite willing to trade with you at current market rates. Both you and they would be able to save railway carriage by this direct dealing.

   Hotel and restaurant proprietors and boarding-house keepers would also often be willing to enter into a contract with you.

3. *The private customer.* This is the best trade to capture when you have only a few carcases to dispose of, as both consumer and producer (*you*) save the middleman's profits. But when a large number of carcases have to be sold the time occupied in selling them is hardly worth the money saved. The aid of a middleman may therefore be invoked.

   The exception is when you have a green-grocery or similar round, when your customers would often be pleased to purchase a fine juicy rabbit from a source which they know is reliable.

Occasionally a local trader or a private customer prefers a rabbit unskinned and undressed. When this happens his likes should be ascertained and studied.

But what, say you, are we who live far from a town to do with our small surplus after we have supplied our fellow villagers' requirements ? Well, there are many men in our villages, as butchers or general shopkeepers, who are always driving about for orders, or to deliver goods, and who could very well make a point of collecting, dressing, and forwarding to the larger centres all the tame rabbits to be obtained on their usual rounds. It would be to their own interest, as well as to the advantage of clients, to see that a suitable kind of animal is bred, and it would pay them to import into the district from time to time bucks for stud purposes. This would improve the breed and weight of the native animals, and save interbreeding with near relations, which so soon deteriorates the fecundity and vigour of any live stock.

If they agreed to take all animals of a suitable weight, and the neighbourhood supported them by supplying all they could, it would soon be found that a considerable sum was being brought into the district by this means, to the advantage of all concerned.

Further, why should not some enterprising breeders start a round similar to that of the egg higgler, purchasing the rabbits from fellow-breeders round about and consigning in bulk to the big markets ?

THE WEIGHTS THE MARKET WANTS

The carcase that is most popular with the London trade is one which, when dressed, weighs from 2½ to 3½ lbs. Larger carcases usually fetch less money in proportion, the difference sometimes being as much as a penny or twopence per pound. In some parts of the provinces, however, between 3 and 4 lbs. is the favoured weight. Breeders should therefore ascertain the popular weight in the market they intend to patronise before sending in a consignment.

This word of caution is necessary ; frequently salesmen receive mixed consignments, the bulk of which are unsaleable in their district, but which would fetch a good price elsewhere. A large poulterer told me the other day that in a hamper he had just received, some specimens weighed over 8 lbs., while others barely turned the scale at 2 lbs.

When a rabbit is to be killed it should first be stunned.

The table breeds named elsewhere should easily reach from 4 to 5 lbs. when twelve or fourteen weeks old. If in good plump condition they will usually lose about one-third of their live weight when skinned and dressed ready for cooking. Thus to cater for a market requiring a 3-lb. rabbit you would have to kill off the youngsters when they weigh about 4½ lbs.

The need for a spring balance in every rabbitry will be apparent when it is realised that the weight of each market rabbit has to be tested.

HOW TO KILL A RABBIT

Rabbits intended for killing should be liberally fed for a week previous, but food should be denied them on the last day. If they are to be killed early in the morning they should have no food the night before. Thus any undigested matter they may be carrying will be cleared out of the system.

The one important point to remember when killing a rabbit is not to bruise the neck any more than can be helped. The killing must not be done by dislocation of the neck, as this would discolour the flesh about the neck, owing to the blood settling there. The Ostend rabbits sent to us by our continental neighbours were always bled, hence their whiteness as compared with the flesh of a wild rabbit. We, too, must bleed our rabbits.

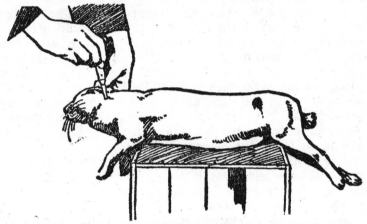

In killing, a sharp-pointed knife should be thrust right through the neck close behind the ears.

In killing, then, hold the rabbit up by its hind legs with your left hand, and with your right give it a sharp blow on the back of its neck just behind the ears, as seen in the picture. This will render it insensible. Even in this way the neck will be somewhat bruised and will spoil the appearance slightly when marketed in the modern fashion, but on humanitarian grounds one has to sacrifice something.

Having stunned the rabbit, hold it, head downwards, over a bucket, and pass a *sharp*-pointed knife right through the neck close behind the ears (*see sketch*). If done properly and quickly the blood will then flow freely, and the rabbit's death will be quite painless. In fact, it will be dead in far less time than it has taken me to describe the operation. I can kill, skin, and dress twelve rabbits in an hour without the slightest difficulty.

HOW TO SKIN A TABLE RABBIT

Next make a slit in the left leg, between the tendon and the bone, and pass the right foot through that slit. Then suspend the rabbit by the hind legs to a hook in the wall.

In opening the rabbit, care must be taken not to break or cut any part of the intestine, as this would necessitate washing, which would naturally spoil the appearance. First, using two fingers of your left hand, take hold of a piece of fur from the centre of the belly, and cut off a small piece, so that you may see the flesh of the flank. Then make a cut in the flesh about 4 inches long, downwards, towards the hindquarters. You will then be able to remove the entrails and stomach with ease. The liver, heart, and lungs must not be removed.

The next step is skinning, and before you tackle this you will be wise to watch the operation in a neighbouring poulterer's if possible. If not, practise on several rabbits to be consumed at home.

Having laid the rabbit on the table, first separate the skin from the flank, then pull the skin, not the flesh, gently, and at the same time work your finger and thumb towards the hindquarters.

With your left hand push the leg through the skin, holding the flesh of the leg with your right hand, give a sharp pull, and the skin will then break away at the hock, leaving the leg free. Go through the same procedure to skin the other leg.

To separate the skin at the tail, give a sharp but firm pull. (Note—The tail is skinned, but about threequarters is left on the carcase.)

Having now skinned the hindquarters, again hang the rabbit up, this time by its left leg, on a hook, a slit having previously been made, and with both hands give a long pull. This will remove the skin as far as the ears. Cut off the ears closely, also cut over the eyes (*the eyes should not be removed*), and continue using the knife until the head is completely skinned.

Chop off not quite half the feet of the hind legs, leaving about 1½ inches of fur on, and pass the right leg through the hole already made in the left leg. Then chop off the paws of the front feet, and tuck the front feet through the first and second rib, as shown in the illustration.

Finally the carcase should be wiped out with weak vinegar and water and laid on its back on a flat surface and on top of it should be placed a weighted board, so that it may be shaped. Shaping will be completed when the carcase is thoroughly cold, and it will then be ready for despatching to market.

A rabbit prepared in the Ostend style.

PACKING THE CARCASES

Before packing, allow the carcase to become quite cold, or it will travel badly. This shaping and cooling is best done in a dark room, as light tends to spoil the white appearance of the flesh.

The carcases may be packed in either hampers or boxes lined with *clean, white* paper, with a sheet of paper between each layer and between the top layer and the lid. I have found that where the rabbits are dressed in the Ostend style, that is, without skins (and this is the only method acceptable to the markets), they travel best when packed in wooden boxes, ventilated in a similar way to foreign egg cases. Suitable boxes can often be purchased at the local stores for a trifling amount. Boxes answer the purpose better than baskets. The latter are generally somewhat expensive, and have to be returned, whereas if cheap boxes are used they can be sent as " non-returnable." In this way the railway charges are lessened. But as the railway carriage charges are based on the actual weight of each package, the breeder, in his own interest, should see that only light boxes are used for packing his supplies.

Care should be exercised to see that the carcases are packed tightly together, so that they will not become loose through the shaking which they receive whilst they are in transit. If they are packed loosely they are apt to get bruised, and any signs of bruising will reduce their value immediately.

Where the consignments are to be sent by rail, they should be forwarded by passenger train and consigned at owner's risk rate. All boxes must be labelled distinctly with the full address of the consignee, together with the name of the nearest railway station. The box or package

This is the sort of box in which to rail live rabbits. If the compartments are lined with hay they will travel without possibility of harm.

should be plainly marked " PERISHABLE," and the sender's name should also be stated, for the benefit of the person to whom they are consigned.

Tie-on labels are not to be recommended, as they are so easily torn off. The labels should either be tacked or gummed on the box. Post-cards make excellent substitutes for labels.

MARKETING LIVE RABBITS

It occasionally happens that a trader prefers to receive his rabbits alive so that he may keep them until required by his customers or kill and dress them in his own fashion.

The one rule to remember when sending live rabbits by rail is not to overcrowd them into a receptacle that is only half large enough for them. If you do the probability is that they will present a very bedraggled appearance upon arrival at their destination, they cannot be graded with others considered to be of tip-top quality, and, consequently, the prices realised will be much less than would have been the case had you appreciated this point when making up the consignment.

The box in which they travel should be clean and lined well with straw or hay, and before being despatched the rabbits should be given a good feed and a good drink.

The special travelling case seen in the sketch is equally suitable for rabbits destined for the table or stock and is largely in use on the Continent. It can easily be knocked together by any handyman for a few pence.

The special methods of skinning rabbits kept for their fur and of marketing the pelts are dealt with in the following chapter.

CHAPTER IX

RABBITS FOR FUR

THE breeding of rabbits for fur is, to-day, *the* most important branch of utility rabbit-keeping. As emphasized elsewhere in this book, the demand for good quality rabbit skins from the right breeds of rabbits is enormous ; moreover it is an increasing demand. The furriers can do with far, far more skins than are at present being produced. Are not nine out of ten of the fur garments sold in the shops for woman's adornment manufactured from rabbit skins, or pelts as they are called in the trade ?

It cannot be too plainly understood by those who are going in for fur rabbit-keeping, however, that only pelts of first-class quality are wanted by the furrier. Inferior skins, poor coloured and with no depth, stained or showing traces of moulting will, if they are bought at all, command only a very low price. Good pelts can readily be sold at anything between four shillings and seven shillings apiece.

THE PELT-BUYING SEASON

Another point of importance, too, is this : It is difficult to sell any sort of pelt, good or bad, at any other time than during the pelt-buying season. This season extends from the beginning of November to the end of March.

Coming now to actual details, the most popular fur breeds, as has before been said, are, first and foremost, the Blue Beveren, followed closely by the Chinchilla, then the Havana, then the White Beveren, the Lilac, the Sitka, and the Argente de Champagne.

The natural colours of these rabbits are the actual colours best liked by the furrier. The pelts do not have to be dyed therefore, a very definite advantage in view of the prejudice existing in many quarters against the wearing of dyed skins. The colour extends to a good depth, too, and the fur has fine wearing qualities.

'WARE STAINED COATS

The housing, feeding, breeding, and rearing of fur rabbits is carried on on precisely the same lines as it is for table rabbits. In both, the aim of the rabbit-keeper is to secure quick growth, to keep the stock in the pink of condition, to avoid those small ills which so pull down the rabbits suffering from them. With fur rabbits, however, it is more than ever important to observe the laws of cleanliness and sanitation. A stained coat is most probably a spoiled coat. One can sometimes remove a stain by rubbing white of egg into the fur, letting it dry and then

The four favourite breeds of fur rabbits.
The Blue Beveren is first favourite to-day.

brushing the spot vigorously, but prevention always was, and always will be, better than cure.

My own plan is to use no actual bedding material for the fur rabbits, no straw or hay in the sleeping compartment, that is. What I do is to sprinkle the floors of both compartments liberally with sawdust, to remove that which is soiled *every day*, and to replace the whole lot once a week, or oftener if necessary. Peat moss litter, well broken up, dry ashes, or even dry earth, anything, in fact, which will absorb the liquid manure, can be used instead of the sawdust, *provided it is never allowed to become sodden*.

The hutches must be exceptionally roomy. Constant friction against a rough wooden wall would be liable to damage the pelt. So, too, of course, would any projecting ends of wire netting, any jagged nails, any splinters.

There must be absolutely nothing which can yank out lumps of hair.

THE NEED FOR SHADE

As the beautiful blue colour of the Blue Beveren is liable to fade in strong sunlight, and the rich brown of the Havana to turn rusty, and as the colours of other of the fur-bearing breeds may be similarly affected—though I have not proved that myself—it is better, when possible, to keep all fur rabbits in an indoor rabbitry. When there is no shed available, however, it would still be an easy enough matter to arrange a sun awning over the front of the hutches.

Only *strong* summer sunlight is harmful, remember. Neither fur rabbits nor any other rabbits will thrive if forced to spend their lives in deep gloom. They may sleep underground in the wild state, but a good deal of the daylight is spent out in the fields.

SOME BREEDING HINTS

Now a few words about the breeding of fur rabbits.

As with all other stock, the young inherit the good and bad qualities of their parents. Only if the doe *and the buck* have a first-class pelt will the litter have pelts of uniform good quality. So in starting a fur rabbitry it is all important to buy really good stock, strong in all the points that go to make a good fur rabbit ; it is all important to save for breeding purposes only those youngsters that are as near perfection as it is possible to get ; it is all important only to mate one's stock when they are in perfect condition. To mate a rabbit that is moulting is to ask for shabby-coated litters that, no matter how careful one may be with them, will never grow out of their defect.

I have said above that the rabbit pelt-buying season commences in November and carries on into the early spring. Obviously, then, one's aim should be to have the litters in perfect coat and ready for killing from November onwards.

The reason why pelts are only wanted between November and March is pretty obvious. Rabbits then have their winter coats, and the winter coats are far fuller, deeper, and better all round in quality than the summer coats.

It would appear, then, that one has only to mate one's

The Flemish Giant, a grand breed for table purposes.

To face page 96.

stock so that they produce litters which have time to assume their winter coat before March is far gone, to make fur rabbit-keeping a success. Yes that is true, to a certain extent. But there are two other factors to be considered:

1. The young must be of good size when they are to be killed in order that the pelt may be a good-sized pelt.

2. Every week one keeps on a youngster adds to its cost of keep, and every penny added to the cost of keep detracts from the price obtained for the pelt.

THE BEST MATING MONTHS

What we have to do in order to get the biggest profit out of our business is to mate our stock so that the young are produced at a time when they will make the quickest growth at the smallest outlay on bought food and yet get through their moult and assume their winter coat just at the time when prices are ruling highest—usually early in the buying season. Winter litters might be expected to have attained a bigger size by November than spring litters, having had another two or three months in which to grow. In my experience, though, a youngster born in March or April will invariably catch up one born in January or February. And the latter will have had to be fed for several weeks longer, and that through a period when pretty well all food must be bought at the highest price of the year !

My own practice is to mate a good many of the does so that they cast their litters in March—when the weather is all in favour of rearing and wild food is beginning to be plentiful—and to continue mating the rest of the does month by month until August, when breeding ceases. The March litters provide fine pelts for disposal in November and December, and the later litters keep the supply going through the winter until the end of the buying season.

It never pays to keep over fur rabbits—unless, of course, they are to be used for breeding—from one season to another. The food they would then eat would cost far more than the price obtained for their pelt and for the carcase.

G

HOW TO TEST FOR MOULT

Generally speaking, then, we must aim to kill off all our pelt rabbits between November and March. But we do not kill them off in the exact order in which they were born. The condition of the coat at the time when we would do our killing is the deciding factor. It may quite likely be that a May-born rabbit is in better coat in January, say, than a March-born rabbit.

The way to test the pelt of a rabbit in order to see whether it bears any trace of the recent moult is this: Take the animal out into strong sunlight, hold it snugly in the arms and peer closely *into* the fur, blowing it apart here and there in order to facilitate the examination. If the coat is in saleable condition it will be of even colour all over and well down towards the skin. It will also be of even texture and depth. The actual skin—and this is most important—will be a pale pink tint all over, without any darker patches. The presence of such dark patches in the case of all rabbits save white ones is a definite indication that the moult is still proceeding ; the patches are, in fact, patches of new fur in the process of forming.

For the benefit of the few who might imagine that one or two small blemishes would go unnoticed by the furrier who was examining a whole parcel of pelts offered to him, let me add that the dark skin patches above referred to are far more obvious on the reverse side of the skin. The reverse side is, of course, available for inspection after the rabbit has been killed and the pelt removed from the carcase. The keenness of the buyer to spot such patches when they exist will be understood when it is realised that the hair from the patches will often fall out during the dressing process, rendering the whole pelt valueless !

PATCHES ON SKIN ARE SIGNS OF MOULT

How to test whether a rabbit has finished the moult or not.

One other point concerning the moult. The rabbit carries its winter coat only during the colder period of the year. Thus, whilst care must be taken at the beginning of the

pelt season to see that the summer coat has entirely
disappeared, equal care must be observed towards the
beginning of spring to see that the animals to be killed are
not starting to shed their winter coat and don their
summer coat. There is danger, you see, in being late in
the killing as well as in being early.

HOW TO KILL A FUR RABBIT

It is quite a simple matter to kill a rabbit. A good
method to follow is fully described in Chapter VIII. I
prefer, however, to do the killing by striking an extra
smart blow on the back of the neck, as depicted in the
illustration on page 88, thus obviating the necessity of
bleeding the animal. The process of bleeding, unless one
is very, very careful, is liable to stain the pelt. True,
blood-stains can be removed if they are sponged off immediately with warm water, but better to have no stains at all.

When dead the rabbit should be hung up, head downwards, for half an hour, and then skinned. To delay the
skinning until the carcase is stone cold is to render the
operation doubly difficult. It is also a mistake to do the
paunching before the skinning. It is practically impossible
to prevent the coat from becoming stained during the
paunching process.

HINTS ON SKINNING

Here is the best method of skinning a fur rabbit:
Setting up a small, firm table in some cool, airy place
out of direct sunlight—and therefore less liable to attract
flies—lay the rabbit on it, on its back, and with a sharp
knife slit the skin in a straight line down the centre of the
belly from chin to tail and up the inside of each leg from
this centre line to the " wrist " and hock respectively.
Cut round the wrist and hock and at the neck behind the
head. Feet and head skins are not wanted on the pelt.

Now, starting at the belly-line, peel off the skin at one
side until the legs are cleared and repeat at the other side.
Next cut off the tail, which is rarely wanted, and, starting at
the stern, peel off the skin and finish at the back of the head.

The next process is the drying of the pelt. This is *most*
important and must be done thoroughly.

By means of drawing-pins or tacks, preferably the former, fasten the pelt firmly to a flat board. Fix it fur side downwards and draw it only sufficiently tightly to smooth out wrinkles and creases. You will find it simpler to fix the two foremost corners first, fixing the sides section by section, as the pelt is smoothed out, and finally fixing the rear ends, that is, the hind legs. Never pull the skin lengthways, only by gripping the sides.

THE IMPORTANCE OF CAREFUL DRYING

Drying is best carried out in the open air, whenever the weather is suitable. We don't often get good drying weather between November and March, however, so more often than not the rabbit-keeper must resort to artificial drying.

Given a favourable spell, then, the board, with the pelt attached, should stand at an open window through which a current of fresh air passes. In the earlier part of the season, when flies are about, it will be necessary to place a single layer of butter muslin over both board and pelt to keep the pests off the latter. Fly maggots would, of course, spoil the pelt. In unfavourable weather, board and pelt should stand in some room where there is a fairly constant temperature —in the kitchen, for instance—though not too close to the fire. On no account must the pelt be placed in the oven.

Twenty-four hours or so after drying has commenced the skin should be scraped, all adhering fat being removed. A blunt knife is the best tool to use, but it must be handled carefully ; to penetrate the skin or to " wear it thin " at any point is to spoil it.

After the scraping, the drying process should be completed, and then the pelt is ready for the furrier. Really it is best despatched straight away, which indicates the advisability of killing a dozen or more rabbits at a time, so that a decent parcel may be made up. Naturally it doesn't pay to send off less than a dozen skins to the furrier at a time. If for some reason or another the pelt has to be kept at home for a bit it should be stored in a dry, airy place, several skins being packed back to back, as it were. A sprinkling of naphthalin might be necessary if pelts are to be stored any length of time, in order to keep moths at bay.

HOME-CURING

It will be noticed that I have not said anything about a curing process. *Rabbit skins intended for the furrier must not on any account be cured in any way.* The furrier will accept them only if they have merely been dried in the way described above. If the skins are wanted for home use though, here is the procedure to follow: ·

Stretch the pelt on a board, as previously advised, scrape it clean of all adhering flesh and dry it with a piece

The proper way to skin a rabbit the pelt of which is destined for the furrier,

of clean linen or cloth. Then every day for at least a week sponge it with the following solution :

Saltpetre	1 ounce
Alum	½ pound
Water	1¼ pints

The best way to make the solution is to boil the water and, whilst it is still boiling, to pour it over the alum, previously placed in another utensil, stirring all the time until the alum has completely dissolved; then add the saltpetre.

Apply the solution to the skin side of the pelt by means of a small sponge. Take the very greatest care that no liquid comes into contact with the actual fur; otherwise it will cause the hair to fall out. After each daily sponging, bring the pelt, still attached to its board, into a room where there is a more or less even temperature and leave it there until the time for the next day's treatment comes round.

When the alum treatment has been completed, detach the pelt from the board, but again hang it up in the same even-temperatured room, where it should remain for another week. By that time the curing process will be complete, though the skin will be very stiff and harsh, especially if the right proportion of saltpetre has not been included in the dressing solution. It prevents a certain amount of stiffness if the skin is worked about with the hands after it receives its daily dressing, but that means that it must be removed from the board, which is a nuisance.

It will usually be found, however, that a gentle rubbing with pumice stone over the skin side for a few days in succession will soon get the pelt back into its original supple state.

ON MARKETING PELTS

Finally a word about marketing pelts—uncured pelts, that is, for once again let me add that cured pelts cannot be sold.

It is best to include at least a dozen pelts in a consignment. The furriers do not like to be bothered with smaller numbers. The pelts in each consignment should be as well matched as possible, for naturally the skins used in the making up of a garment must be of precisely the same hue, texture, and quality. When all the skins in a parcel match, the furrier has not to go to all the bother of finding skins from another parcel to pair with them.

Those who run an extensive rabbitry should always make a point of grading the pelts they have to sell. Two grades are usually sufficient, skins which are of good size, well furred and well coloured, being included as first grade and all the rest as second grade. The first grade will naturally realise a better price than the second grade.

The pelts should never be doubled or creased when being sent off ; the box in which they are to be consigned should be large enough to permit all the skins to be spread out flat. They should be packed fur to fur, in pairs, the package being made waterproof.

The Editor of " The Smallholder," 18 Henrietta Street, Covent Garden, London, will be only too pleased to supply any reader of this book with the names and addresses of furriers to whom parcels of pelts can be disposed of at current market rates.

CHAPTER X

AILMENTS AND DISEASES

ALTHOUGH rabbits are not altogether immune from various ailments and diseases, there is probably no other kind of live stock less liable to sickness if only ordinary and common-sense methods of treatment are adhered to. Readers will have gathered from the foregoing chapters what these methods are, but I may sum up here a few of the things which have a direct bearing on the health of the inmates of a rabbitry :

1. *The hutches.* While almost any box can be made into a serviceable hutch you must always bear in mind that unless it is draught-proof, damp-proof, and kept scrupulously clean it is *not* serviceable, but, on the contrary, is nothing more nor less than a death-trap. Even with every attention bestowed upon it, it is wise to use a little disinfectant upon it now and again, to give it a thorough cleaning out periodically (apart from the usual bi-weekly clean), and to lime-wash it, or at least wash it well with disinfectant, once a year.

A hutch that has housed a rabbit suffering from some infectious disease should on no account be used again until it has been thoroughly soaked with disinfectant and stood out in the sun for at least a week.

2. *Ventilation.* Many, many rabbitries (indoor rabbitries I refer to) suffer from poor ventilation, and provision for allowing a current of air to flow *right round* the hutches is seldom made. The best way to secure this is not to have the hutch flush up against the wall. In the hutches themselves it is a wise plan to bore a few holes right at the top to allow foul air to escape from them.

Another common fault with rabbitries is that no regard is paid to the supply of fresh air within them. No matter how well-ventilated and nicely arranged your hutches are, it is of little use if the shed in which they stand is poorly ventilated. One usually sees the door made to fit close without any regard to fresh air. This matters little where there is a window or roof ventilator, but if not, a space of about 2 or 3 inches should be left at the *top* of the shed door. It will make a wonderful difference, and instead of there being a heavy atmosphere inside, the air is sweet and fresh.

In cases where door ventilation only is adopted, however, one must arrange the hutches so that currents of air do not beat down directly on to the rabbits. The hutches—those facing the door, at any rate—should certainly be below the aperture in the door.

It follows that if the rabbits have plenty of breathing space they have a natural chance to keep healthy, and it requires only the simplest of rules to be adopted to bring this about.

These remarks as to ventilation naturally only apply to indoor rabbitries; outdoor rabbitries fortunately have their fill of fresh air. The same rule as to not having the hutch flush against the wall still applies, however.

3. *Coddling.* Instances frequently occur where from a too enthusiastic desire to keep the rabbits warm during inclement weather they are very often nearly suffocated. Sacking or boards are so placed as to shut out all means of ventilation, and this is a great mistake. Any device used for protecting hutches standing in the open air need only be loosely hung or fitted to the hutches. They are only necessary to prevent the driving rain or sleet from being blown into the hutch and not to shut out fresh air.

Rabbits do not want coddling. They are naturally hardy, and if kept *dry* they will rarely, if ever, suffer ill-effects from cold weather. Let those who use shutters, therefore, see to it that plenty of air holes are bored into them if they want their stocks to be healthy.

4. *Overcrowding.* Perhaps the most important point to realise, as bearing directly on health, is not to overcrowd. If you have not unlimited room—and very few people have —limit your stock accordingly. Little duties that are pleasures become exceedingly irksome if you find you

cannot move about with ease to attend to the wants of the
stock, and if hutches are in all manner of out-of-the-way
places they are bound sooner or later to suffer from in-
attention.

Give the rabbits plenty of breathing space and ventilation ; then at least they have a chance to keep healthy, but
when you overcrowd you immediately seek trouble of a
serious kind.

5. *Watchfulness*. Always be prepared to tackle any
ailment the moment the first symptoms are noticed, and
always be on the look out for the first symptoms. A
medicine cupboard, equipped with all the simpler remedies, a sponge, a lance, a small enamelled bowl, a bottle
of vaseline, a small brush, a liquid measure, a small
but accurate pair of scales, flowers of sulphur, a box of
ointment of a known healing value, etc., should be found
in every rabbitry.

The majority of ailments and diseases have preliminary
symptoms which cannot escape the notice of a watchful
attendant. If they are taken in hand at once, and the
patient isolated right away from the other stock, a quick and
easy cure is in all probability possible. If allowed to pro-
ceed unchecked they may not only result in the death of
the particular patient, but may infect the whole of the
stock.

Generally speaking, you can reckon that a rabbit which is
bright of eye, alert, keen for its food, has a smooth, sleek,
and good conditioned coat and whose droppings are firm,
is in a perfect state of health, while one that is listless or
mopy or appears sleepy during most of a day and indifferent
to what is going on or to its food, or whose coat is rough
and unkempt, is ailing in some way or another and at least
needs specially watching.

Sometimes such mopishness is akin to our own " liverish-
ness," and in that case a dose of flowers of sulphur will put
matters right. Indeed, believing that prevention is better
than cure, I regularly give my stock a dose of this useful
medicine in a meal mash after a sudden change of weather
or when any other little thing has occurred liable to upset
them.

Here now are the diseases and ailments which rabbit-keepers are likely to come up against.

Abscesses

Symptoms. A hard lump, usually hot and very tender to the touch and growing softer as it matures.

Remedy. Wait until the soft stage is reached, then lance the lump (steeping the lance in boiling water before using it) and squeeze out the matter it contains with a firm but gentle pressure. Get an assistant to hold the rabbit as shown on page 75, while performing the operation. After operating bathe the wound with warm permanganate. Keep the wound open for a couple of days and trim the hair around it to prevent it matting with the discharge. Give the patient a dose of flowers of sulphur and coax it to eat by offering favourite tit-bits.

Colds (*infectious and contagious*)

Symptoms. Sneezing, a running from the eyes and nose, and inflammation of the eyes. Caused by draughty or damp hutches, or bad ventilation.

Remedy. Keep the patient warm, avoid draughts, give a feed or two of warm milk. When the eyes are affected, bathe them once a day with the following lotion :

    Sulphate of Zinc ..   .. 2 gr.
    Water   ..   .. 1 oz.

Coughs (*infectious*)

Symptoms. Mopishness, accompanied by coughing more or less severe. Often the symptoms of cold.

Remedy. As for colds. If severe, add a few drops (not more than six) of sweet spirit of nitre to the milk.

Colic

Symptoms. Uneasiness, obvious pain, and a swollen body. Caused by unsuitable food.

Remedy. Give a teaspoonful of castor oil and rub the stomach gently until the patient is relieved. The most convenient way to give the castor oil and, in fact, all medicines is to hold the rabbit firmly between the thighs, press its head well back and insert a spoon containing the dose between the teeth at the side of the mouth.

Constipation

Symptoms. The sufferer will sit quietly in a corner and will ignore food. Its body assumes a distended appearance.

Remedy. The recognition of water as a desirable drink for rabbits has considerably diminished troublesome cases of constipation, and although the most simple, the water cure is as effective as anything I know If it fails, give a meal mash with which has been mixed a small amount of flowers of sulphur, or in very severe cases a teaspoonful of castor oil. Rabbits fed plentifully on good green food, regularly given water to drink and kept in hutches large enough to allow them to move about will seldom suffer from constipation.

Diarrhœa

See Scours.

Ear Canker (contagious)

Symptoms. A slight or heavy yellow discharge from one or both ears, often loss of condition and weight, and great pain (sometimes denoted by squeals) when the rabbit is handled by the ears. The disease is caused by dirt and dust accumulating in the base of the ear, and the presence of a minute parasite which multiplies quickly. If it is allowed to go unmolested for any length of time, the inside of the ear, extending from the base to a good half-way up, will become encrusted with a dry or sometimes wet discharge, which will eventually penetrate through the ear.

Remedy. A large number of breeders in treating for canker first syringe the ear with soapy water to soften the discharge, remove as much of the discharge as possible without hurting the rabbit, then dust the ear with flowers of sulphur; repeating the treatment daily until a cure is effected.

I do not favour the above method, however, because in the first place it is troublesome, and in the second place, it does not always effect a *permanent* cure. I follow the advice of a big breeder, who has been remarkably successful in employing a strong disinfectant capable of destroying the parasite—carbolic ointment at the strength

of one in twenty-five. He has invariably found two applications to be sufficient. Afterwards, the rabbit is fed
generously on a variety of good foods for a few days.

Ear Wax

Symptoms. The ears have a scurfy appearance, and
there is often a collection of wax low down.

Remedy. Remove the wax with some blunt instrument,
such as a piece of stick, if necessary softening the wax
first with warm soapy water. Be as careful as possible not
to make the ears bleed, as they are very tender. It will
be found convenient to place the patients on a table and
let an assistant hold them firmly by the skin of their
backs. Afterwards sprinkle into the ear, as far down as
possible, a little flowers of sulphur. Repeat two or three
times at three-day intervals.

Ear wax is a forerunner of ear canker, and the ears of
all rabbits should be frequently examined for signs of it.
It is a very simple matter on two occasions in the week—
say, after the evening meal has been given—to examine
one's rabbits and thereby nip in the bud any trouble that
may, if neglected, prove serious.

Eczema (contagious)

Symptoms. A grey scaly skin can be *felt* at the roots
of the hair round the eyelids, nose, and ears, but sometimes
cannot be *seen.* The hair comes away slightly. It is
caused by a lack of green food and badly ventilated and
dirty hutches.

Remedy. As for scurf.

Fits

Symptoms. An animal in a fit has a wild or glassy appearance about the eyes, and will fall over on its side, at the
same time kicking and struggling. The neck will appear to
be stiff, and will be thrown right back, and difficulty will
be experienced in breathing. Fits are brought about
by a bad attack of indigestion, caused through improper
feeding. They are more frequent in young than in adult
stock.

Remedy. Give a small dose of brandy, diluted to half
its usual strength, and repeat every three hours, until the

condition is normal. Afterwards give a regular supply of green food with good herby meadow hay, and oats.

Festered Wounds

Symptoms. Wounds caused by fighting, etc., become inflamed and frequently discharge. Caused by dirt, etc., getting in the wound or very low condition.

Remedy. Lance the wound (*see Abscesses*), squeeze out all matter, wash with Condy's Fluid and apply a dressing of carbolic ointment frequently.

Insect Pests

Symptoms. While neither an ailment or a disease, the presence of insect pests predisposes a rabbit to both. Rabbits attacked are dull and grow thin. Fleas and other pests will usually congregate round the ears and about the forehead. Insects will only be found on rabbits housed in dirty, damp hutches or where mouldy hay or straw is used.

Remedy. Rub flowers of sulphur well into the fur and also dust the corners of the hutch with the sulphur.

Liver Disease (*highly contagious*)

Symptoms. Young stock suffering from a diseased liver show the following symptoms : Eyes very bright with a glassy appearance, ravenous appetite, condition poor, backbone prominent, belly large. Youngsters from about six weeks to twelve weeks old, when suffering from the complaint, usually die somewhat suddenly. If any rabbit-keeper should have a litter of youngsters that do not appear to be growing as fast as they should do, my advice to him is to kill the smallest one, and examine the liver very closely for any small white or sometimes slightly yellow spots scattered about as in the illustration on page 110. If, on examination, these spots should be present, then the best and also the cheapest remedy in the long run is to kill the whole of the litter right off, for where one is affected so is the whole litter, as a rule.

There are two stages, the mild and advanced. If attacked in the mild form, the animal is stunted in its growth and is slow in reaching maturity, while the flesh will never be found to be as firm as with a healthy specimen. At an advanced stage of the disease the stock, as before stated, die off suddenly at an early age. They eat well and appear fit right up to the last.

Liver disease is *the* worst enemy the rabbit-keeper has to contend with, for it is found in all classes of rabbits at all ages. It is highly contagious and is also inherited from the parents. The result is that it is growing steadily, and unless drastic measures are taken it may soon cause almost as much havoc in our rabbitries as Isle of Wight disease has caused among bees.

All hutches that have housed a rabbit suffering from dis-eased liver should be thoroughly disinfected and given two good applications of lime-wash. All litter and other rubbish in and around the hutch should be burnt and the carcase of the animal should be buried.

Remedy. There is no known cure for liver disease, although I have occasionally heard of sup-posed cures. The only "remedy" is prevention. Never breed from anything but perfectly healthy and mature stock, avoid in-breed-ing, never allow a rabbit to eat off the floor of its hutch, always from a trough, always feed care-

A bad attack of liver disease, or spotted liver.

fully and regularly on clean, wholesome food, keep hutches perfectly clean, and well ventilated and damp-proof, and kill off any sufferer immediately liver disease is suspected, and when it is confirmed kill off all that have been housed with that particular animal.

Rabbits that have died from liver disease are not fit for human food. Those that have been suspected of the disease and have been killed may be eaten if the disease is not in a very advanced stage.

Mange (contagious)

Symptoms. Closely akin to Eczema (which see), only the hair falls out freely.

Remedy. As for Scurf.

Ophthalmia (contagious)

Symptoms. The eyes appear sore and swollen, and if examined closely little red pimples will be noticed around the rims and also about the forehead and ears. There is great difficulty in opening the eyes and sometimes they remain closed for several days. This trouble is caused by allowing the hutch to become soaked with urine. It is most frequent in badly ventilated hutches. It may lead to permanent blindness if not taken in hand at once.

Remedy. Transfer the patient to a clean, well-ventilated hutch, free from draughts and lined with sawdust. In mild cases bathe the eyes twice a day for several days in succession with warm milk and water ; in serious cases bathe with this lotion :

Boracic Acid	..	..	..	..	1 scruple.
Water	..	..	..	..	6 ozs.

Feed the patient liberally with green food.

Paralysis.

Symptoms. The rabbit seems to be powerless and drags its hind legs about. It is a painful and distressing disease, and affects both young and adults. The trouble is generally brought about by cold and damp hutches ; and, at times, by improper feeding. It has become very prevalent of late and accounts for thousands of promising youngsters in a twelvemonth.

A cure can usually be effected in the case of young rabbits provided immediate steps are taken. Rub the loins and hindquarters of the patient with turpentine regularly once or twice a day, and give every other day one small pill of the following :

Sulphate of Iron	..	..	..	2½ grains.
Powdered Camphor	..	..	..	4 ,,

Mix these into a little treacle and roll into small pills.

The rabbit will usually eat well during the first stages of the trouble, but will afterwards go off its feed owing to want of exercise.

Pot-Belly

Symptoms. The body is swollen in the lower part. The patient's appetite is normal for a time, but gradually lessens. Lack of exercise, an excess of wet or very juicy green food, or a very hearty meal of some favourite food may all cause the trouble. It, however, usually only attacks rabbits between the ages of two and six months.

Remedy. Give the patient a good run, if possible in a shed, house in a roomy hutch, and feed on hay and whole oats and plenty of water. When the trouble is first noticed give the patient a teaspoonful of castor oil before the first feed in the morning.

Redwater

Symptoms. The urine seems as if tinged with blood and the patient is mopish, though usually feeds well. Dampness, exposure, and improper feeding will all induce this disease.

Remedy. House warmly, feed well, give barley water (made in the usual household way) to drink, with four drops of sweet spirits of nitre every other morning until the trouble disappears.

Scours (*Diarrhœa*)

Symptoms. The droppings are loose and watery instead of being in small, hard balls. The rabbit looks thoroughly ill and speedily becomes thin, unless cured. Too much green food or wet or stale green food are predisposing causes. Young rabbits are the most common sufferers.

Remedy. Stop giving green food and only feed dry food. An acorn powdered and mixed with oats will do good. A dish of cold water will often effect a cure without any other treatment, especially if the patient is not accustomed to it. A little arrowroot may with advantage be added to the drinking water after the first drink has been taken. In very obstinate cases, five or six grains of precipitated chalk will invariably effect a speedy cure. When the droppings have again become normal go back to the ordinary feeding only very gradually.

Scurf (*contagious*)

Symptoms. The patient's skin is rough and flakes off. It is first to be noticed on the nose and eyelids and round the ears. Overcrowding and dirty and ill-ventilated hutches are the main cause.

Remedy. Rub the affected parts well with

Lard	**2 parts.**
Petroleum or Sulphur	**1 part.**

or—

Flowers of Sulphur	**1 part.**
Olive Oil	**6 parts.**

Apply thinly and rub well in every third day. A month may be occupied in effecting a cure. House warmly and cleanly and feed well.

Surfeit (contagious)[1]

Symptoms. Similar to Scurf (which see), but in addition inflamed patches from which matter oozes will be frequently noticed. The hair comes away freely. Dirt and bad ventilation are the chief causes.

Remedy. Wash the affected parts well with soap and water and apply the following lotion, morning and evening :

Potassium Carbonate	1 teaspoonful.
Water	½ pint.

or a strong solution of Condy's Fluid, though the latter is not so effective. Three times a week add a teaspoonful of flowers of sulphur to a morning mash.

Skin Eruptions (contagious)[1]

Symptoms. What appears at first like a white scurf develops into small sores. Poor blood, due to improper feeding and bad housing, are the cause.

Remedy. Wash the sore places with disinfectant soaps and water, and anoint with

Lard	**2 parts.**
Petroleum	**1 part.**

every other day until a cure is effected. Give half a teaspoonful of flowers of sulphur with the morning feed every third or fourth day.

Slobbers

Symptoms. A running at the mouth which occasionally proves fatal.

[1] Wash the hands thoroughly immediately after handling patients suffering from eczema, and all skin diseases, using a disinfectant soap.

H

Remedy. Wash the mouth three times a day with alum water, give a dose of flowers of sulphur every third morning, and plenty of green food at all times. Don't coddle the patient, but keep it away from draughts and damp.

Snuffles (*highly infectious and contagious*)

Symptoms. The symptoms are fairly easy to recognise. A rabbit so affected will during the early stages be noticed to be continually sneezing, and in the later stage the breathing will become audible and laboured. In most cases there is a heavy discharge at the nostrils, the attacks of sneezing releasing the discharge. The sufferer will have a poor appetite and commence to lose flesh, and unless treatment is given quickly fatal results will soon follow.

This is one of the most troublesome rabbit ailments. Stock that have been heavily inbred are the worst sufferers. Badly ventilated and damp hutches and exposure to cold or damp are contributing causes.

Remedy. Keep the patient as snug and warm as possible and feed very lavishly, adding half a teaspoonful of sulphur and a pinch of salt to the morning mash. The appetite may be poor. Always have a dish of clean water in the hutch. Bathe the mouth, eyes, and nose frequently with warm water and vinegar in equal parts.

Three or four times a day until a cure is effected place the patient in a box, just large enough to enable it to move about freely, the floor of which is covered with sawdust and sprinkled with a few drops of eucalyptus. Have a lid to the box and drill a few holes round the sides for ventilation. Leave the rabbit in the box for an hour at a time.

This is much better than the usual method of sprinkling the hutch floor with eucalyptus.

Unless the greatest care is taken the ailment turns to inflammation of the lungs, when the patient invariably dies.

Sore Hocks

Symptoms. Sores appear on the hocks, often extending from the hock to the foot. The trouble is caused frequently by an accumulation of filth on the floor of the hutch and occasionally by impure and overheated blood, due to improper feeding. The sores are made all the worse by the fact that the whole weight of the rabbit's body is concentrated on the hocks.

Remedy. First clean the hutch thoroughly, then put a liberal coating of sawdust on the floor, paying special attention to the corners, and plenty of clean bedding material in the sleeping compartment. Wash the hocks with warm water, dress them with some healing ointment, and bandage them. Renew the dressing and bandages whenever the latter are displaced. Give a dose of flowers of sulphur with the morning feed. If the sores are very bad, to attempt to cure them is a tedious performance; it is most likely to occupy quite six months. In the circumstances, therefore, it is usually advisable to kill off the sufferers.

Sore Mouth

Symptoms. A curious hesitation in commencing to eat.

Remedy. If the teeth are too long file them or nip them with wire-cutters. If the gums are sore wash with alum water.

Sudden Death

A common occurrence among young stock in the spring and early autumn. The rabbits seem to die off without suffering from any disease. They usually eat well right up to the last meal and then a few hours afterwards are found dead. The first, and, to my mind, the most likely, cause is a sudden change of diet. When the summer has been an exceptionally fine and dry one, green food gathered at all times is quite fit to feed to the youngsters without any drying process. Then, perhaps, comes a period of heavy dews both morning and night, together with a muggy atmosphere during the day, which never allows the green food to become dry, and this, together with tinges of frost, are, in my opinion, the cause of some breeders occasionally losing so many youngsters.

There is also the question of poisoning, especially when green food gathered from the roadside is fed liberally. Many County Authorities now supply weed-killers of various kinds to their roadmen, and it would be a very easy matter to poison a stock of rabbits with green food gathered after being subjected to such conditions.

Always gather green food carefully, and when changing from one season's food to another, make the change gradually. After heavy rains or dews, wipe freshly gathered greenstuff partially dry before feeding it.

Vent Disease (contagious)

Symptoms. The vent of the patient is swollen and very much inflamed. It is quite a common complaint, brought about by not allowing rabbits to mate when so inclined. No rabbit suffering from this disease should be mated until a cure has been effected.

Remedy. Dress the affected parts daily with carbolic oil. Keep the hutch scrupulously clean and feed on non-heating foods.

Worms (contagious)

Symptoms. The presence of worms in the excreta and a thin, poor appearance of the patient combined with a ravenous appetite. The trouble is declared to be hereditary, so that no animal suffering from worms should be bred from until cured.

Remedy. Give the patient on an empty stomach from four to six grains of areca nut in a little milk followed at a short interval by a small teaspoonful of castor oil. Repeat the following day if all the worms have not been expelled.

Overgrown Claws

Hutch rabbits cannot keep their claws well worn down by scratching like wild ones can. At times, therefore, the claws become overgrown, and require to be shortened back judiciously, taking care not to cut into the quick.

This can best be done by the aid of a pair of small cutting-pliers.

Moulting

Moulting is not, perhaps, strictly speaking, an ailment, but it is just as well to treat it as such. A moulting rabbit should be groomed daily with a bristle brush and with the hand, and should be given, morning and night, in addition to plenty of hay and greenstuff or roots, a mash made of equal parts of broad bran, oatmeal, and linseed meal.

Some Simple Remedies

Slightly tinting the water with sulphate of iron is a good tonic for an animal a bit out of sorts, and a few holly leaves daily are also a great help in the same direction.

A piece of rock salt in the hutch, always available for the rabbits to lick when so inclined, helps to prevent worms and also tones up the system.

Dandelions make a good blood purifier, groundsel helps

A DROP OR TWO OF EUCALYPTUS OIL - RUBBED DOWN NOSE OF RABBIT. TO CURE WHEEZING

SPONGE OUT WITH CLEAN LUKE WARM WATER - THEN DRY WITH A SOFT CLOTH.

CUTTING PLIERS - OVERGROWN CLAWS -

- DUST FREELY WITH BORACIC POWDER. - EAR CANKER.

A piece of rock salt in the hutch, always available for the rabbits to lick when so inclined, helps to prevent worms and also tones up system.

on the moult, and shepherds' purse is good in cases of scours. It is always wise to resort to natural remedies of this nature when they are available.

How to give Medicines

When handling a rabbit in order to give it medicine, or when it is necessary to attend to its ears or claws, you want to watch out. A struggling rabbit can inflict quite a nasty wound on one's wrists or hands, especially if its hind feet have free play.

If, however, you wrap the animal up in a piece of sacking, or other suitable material, so that only its head projects, the rest is comparatively easy. Also, if you can get someone to hold the patient carefully for you it is an advantage.

. Some medicines can be given to rabbits in their mash. For instance, a pinch of both flowers of sulphur and magnesia can be mixed in the mash for a few mornings and evenings, helping to put right minor digestive ailments. The rabbits will readily take the tonic this way.

Magnesia can be given to a pregnant doe, or to quite tiny youngsters like this, too.

The safest scheme when you have to dose a rabbit,

A bone mustard-spoon is very useful for giving powders direct to rabbits. For liquid medicines use a small syringe for preference, though a fountain-pen filler can be used, if you take care not to break it.

The Rabbit-keeper's Medicine Cupboard

This chapter would not be complete without some reference to the medicine cupboard which *should* be part and parcel of every rabbit-keeper's outfit.

One of the most important items to include in the cupboard is a jar of vaseline. I keep jars of both the ordinary and the carbolised sorts. A jar of boracic ointment is also essential. Vaseline and boracic are useful for the treatment of wounds, which are by no means uncommon even in the best regulated rabbitry. Does, for instance, will assault their youngsters when driving them into the nest box out of sight. Also some rabbits are given to fighting, and will tear each other badly with tooth and claws.

Incidentally, it should be remembered that rabbits which are liable to fight should never be run together.

Permanganate of potash crystals again are of great use, for they are a fine disinfectant. They are also a remedy against scours. For the latter you put a few crystals in warm water and with the resulting liquid wash well the soiled part of the rabbit and dry ; this in addition to the treatment of diet and medicine recommended on page 112. To my way of thinking a drink of fresh water is also helpful in a case of scours. If fresh water is always supplied, many a rabbit will cure an attack of scours itself before the complaint makes headway.

Linseed oil—the refined quality obtained from the chemist—should also have a place in the cupboard. It is of value for temporary digestive disturbances. Half to one teaspoonful for each rabbit should set things right.

Bismuth powder is also a good remedy for digestive troubles. As much as will lie on a sixpence in a small teaspoonful of water is a usual dose.

Cotton wool for swabs should be in the rabbitry cupboard.

Eye trouble is not infrequent, and a bottle of saturated solution of boracic powder should be at hand to bathe the eyes. It can be diluted with a little warm water if desired.

Eucalyptus oil, too, must be to hand for the first symptom of cold. A smear on the nose and a little dropped on a rag which is tied to the wire of the hutch may nip trouble in the bud !

If these remedies are kept in the rabbitry, and are used without delay when required, it saves a lot of worry.

CHAPTER XI

SOME POINTS FOR THE FANCIER

As I have indicated in various other chapters, there is always the possibility of one or more of a utility rabbit's litter being found to possess excellent exhibition points, and when this happens the breeder is a lucky man. Not only may he win prizes at a show but he may either keep the rabbit (if a doe) himself, mate it to a pedigree animal, rear the progeny and sell them for stock at good prices, or (if a buck) let it out at a fee for stud purposes. In this way the profits are considerably augmented.

When such a rabbit is discovered it should be given special attention in the way of grooming, feeding, etc., almost from the first. Then the final preparations just before the show will not be nearly so laborious.

One of, if not *the*, most important features an exhibition rabbit must possess to be successful is condition. No matter how classically it has been bred, how many cups its parents have won, or how good in markings and colour the specimen may be, if it lacks condition it is severely handicapped. Having an open coat, being too fat, or dirty, are penalties which total up negative points or failures.

The coat must be soft and glossy, the flesh firm, the eye bright. There must be such a jaunty air and vivaciousness about the rabbit that even a casual observer will instantly note it has qualities above the ordinary specimens.

To attain this stage good food must be given. Regularity and consistency must be valued. Inferior feeding stuffs must not be used, as in the long run it will be found that what was considered economical will prove otherwise. Two meals a day will be found to answer well in the case of an exhibition rabbit, as too many feeds will cause the

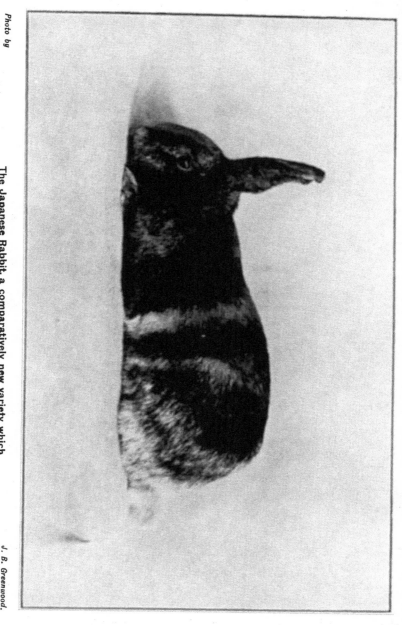

Photo by

To face page 120.

The Japanese Rabbit, a comparatively new variety which
is likely soon to come into great favour.

J. B. Greenwood.

rabbits to get fat and baggy in coat, neither of which is desirable. For breakfast sweet hay, and plenty of green food, or roots, in season, and at night two handfuls of whole oats and wheat (two parts of the former to one of the latter) will be found to be eaten up readily. These are the rations for one rabbit.

It is important not to overfeed, inasmuch as food left untouched goes stale.

A drink of milk may also be given daily, but any left over should be removed, otherwise it often gets overturned and is liable to injure the coat.

This treatment will answer well up to a few days before the show. For the last week give a meal mash made of barley meal and bran, or patent rabbit food and bran, made into a crumbly consistency with boiling water. A piece as big as a duck's egg will be sufficient for one rabbit at a meal.

At night give unlimited green food and a handful of hay, the latter to counteract any probable stomach troubles. Meal mashes and milk are fattening, so they must be used with a sparing hand. On the final day a further variation of oats and wheat for breakfast and green food at night may be given.

The hutches must have regular attention. They must be of a common-sense size, free from draughts, and kept clean. Plenty of sawdust must be used and all bedding material replenished twice a week, whilst the corners mostly used by the rabbits should be cleaned out every day.

If the rabbit happens to start moulting near the date of the show, an additional meal mash or two in lieu of oats will assist it to obtain its new coat quickly and cleanly. A packet of Thorley's Food mixed with the mash improves it, and is a good appetiser, being always relished. A tablespoonful will be found to be sufficient for mixing with the food given to three rabbits, and a ball of mash as large as a good-sized duck's egg usually suffices for a single animal. An addition of boiled linseed will add lustre to the new fur. This is made by putting about a teacupful of linseed into a pint of water, allowing the whole to simmer until it reaches boiling-point, and then pouring it over the meal.

THE FINISHING TOUCHES

During the last week every available minute should be spent in grooming the animal. Grooming removes all loose hairs and puts a gloss on the coat that makes even a second-rate rabbit show up to advantage. A rabbit with excellent points but badly groomed will often lose to a rabbit of inferior points but in perfect condition.

Some fanciers use a silk handkerchief for grooming their rabbits, but I am of opinion that the bare hand is better. Stroke the rabbit briskly from the back of the head to the tail (not forgetting the sides), but don't use too much pressure or you may draw out some of the fur and instead of improving the appearance will cause it to look patchy.

If there are many loose hairs you may use a brush with long bristles or, better still, put just a spot of glycerine in the palm of your hand and go over the whole of the fur carefully. This will usually bring them all away.

WHAT SHOW PRIZES MEAN TO THE EXHIBITOR

There are two benefits, apart from the pleasure, which an exhibitor derives from a success at a show : (1) The prize money ; (2) the publicity and the increased demand and prices he receives for his stock, and of the two the latter is the most important, and, in fact, is really the only compensation to be looked for. More often than not the expenses of showing the rabbit run away with practically all the prize money.

The chief items of expense connected with showing are the entrance fees charged and the railway charges for the carriage of stock to and from the show. The former is generally fixed by show committees at a few shillings for each entry ; but there are several shows where a class is provided for cottagers. In those cases the fee charged is usually a nominal one.

The prize money at the leading shows is awarded to the winners of the first, second, and third prizes, and the amount generally paid is £1, 10s., and 5s. respectively. This amount is small when compared with the outlay.

Nearly all agricultural shows will be found to provide

classes for rabbits.   In addition, there are fanciers' societies in the majority of large towns, and shows are periodically held.

The usual plan adopted is to write to the secretary of the show at which you intend to exhibit, asking for a schedule and entry form.   In filling in the particulars required care should be exercised that you give the correct description of your rabbit, its age and sex, and also the selling price.   Afterwards these particulars should be forwarded by post together with a postal order to cover the amount of the fees.

The secretary, in return, will forward a label, on which is printed the address of the show, also giving the number of the class and that of the exhibit.   On the back of the label should be written the address to which the rabbit is to be returned.

If the show is close at hand, the exhibit is best personally conveyed there and so one item of expense is saved.   But whether near or far, make quite sure that the specimen arrives in good time for the judging.

The travelling-box should be large enough to allow the animal room to turn round, and must be provided with adequate ventilation.   In the bottom of the box should be placed a layer of sawdust, together with a supply of hay, a small quantity of green food and a piece of stale bread.   If several rabbits are being sent the type of box illustrated on page 92 will serve admirably.

On no account must the box be nailed or tied with rope. Secure it with a stout strap and buckle so that it may be easily opened and repacked.

SOME BREEDING POINTS FOR FANCIERS

The breeding of purely exhibition rabbits varies very little from the breeding of utility rabbits, though there are one or two special considerations to bear in mind.   For instance, before a pair are mated the show points of each should be very carefully considered.

You must never pair together two rabbits possessing similar failings.   If you have a doe with exceptionally good head properties it is wise to mate her to a buck with equally good body properties, and so on. The benefits of the observance of this rule must in the long run be evident. It is an easy matter to breed youngsters faulty in markings shape, etc., but a totally different business to eradicate them, and bearing in mind that " like begets like " too much attention cannot be paid to the business of pairing up specimens which, after having weighed up the pros and cons, seem best suited to each other.

It is customary with fanciers to weed out all poorer specimens as soon after birth as possible, and either to fatten them up for the table or drown them forthwith. This allows the doe to give undivided attention to the pick of the bunch.

But do not be in too great a hurry to thin out a litter, because with many varieties the colour and markings are misleading at the outset and do not develop for several weeks. For instance, Silver Greys are black when they are born. It is not until they are a month old that the silvering makes itself manifest. Progress can then be observed day by day, and any that have white patches about them or remain a solid black on any one part can be considered the least valuable to keep.

The value of English rabbits, on the contrary, can be estimated very plainly as soon as the nest is examined.

Those with the spots correctly placed, and having good saddles and clear heads, are the most desirable for retention as breeding stock. The self-blacks which often appear in the litters of this breed can be kept with those badly marked and considered solely for marketing in a dead state.

The same remarks are applicable as regards Dutch rabbits, and it is a very easy matter to separate the good from the indifferent even when they are a day old.

Belgian Hares, again, often look most unpromising until a month old, and then what was thought to be the poorest may turn out a perfect treasure. Generally speaking, however, those specimens that are the darkest colour along their backs turn out the best coloured adults. Any that have brindled fore legs or patches of white on the hind legs are useless, as are also those that are a slate-grey colour on the haunches, and they can be put in the fattening pen the moment they are discovered.

With Flemish Giants we want size first, and in contradistinction to the Belgian Hare, which is of a racy appearance, we want big, square, heavily built giant youngsters, if they are to turn out trumps. Weed out those with narrow heads and backs, thin limbs, and narrow chests. It is easy to pick out a promising Flemish, as far as size is concerned, at a month old. Have in view a heavy carthorse, and build up your idea of what a Flemish should be on those lines. A great fault with this breed is lopping ears, and it seems to be on the increase. This, in my opinion, is due mainly to low and badly ventilated hutches, and should not be encouraged.

THE NEED FOR NEW BREEDS

The number of varieties of exhibition rabbits best known in this country to choose from is sixteen. These are: The English (in black and white, blue and white, and grey and white), the Lop (in numerous colours), Belgian Hare, Flemish Giant, Silver Grey, Silver Fawn, Silver Brown, Dutch (black, blue, steel grey, and tortoise-shell), Himalayan, Black and Tan, Blue and Tan, Polish Angora, Beveren, Havana, and Imperials.

There is no need to go into great detail with regard to each of these breeds, but I have a few words to say about the most popular.

THE ENGLISH

The English, apart from Flemish Giants and Belgian Hares, I think, takes pride of place. It is an exceptionally hardy rabbit to which scarcely any treatment comes amiss. Given cleanliness and good food, this breed will stand many hard knocks. It was long bred without any particular system of points, but of late years a strong club has taken it in hand, and a standard has been fixed, which is rather hard to breed to, many self-coloured specimens being bred from stock that has been carefully mated for generations. It is becoming more popular every year, and if anyone wishes to try a breed which offers great possibilities for the exercise of care and skill in mating, let him try the English.

It weighs from 6 lbs. to 8 lbs. when adult, and is a good fleshed rabbit. The does are good mothers, and very prolific, while the young leave the nest and commence feeding earlier than most other breeds. Recent attempts made to use the English for table purposes have been most successful, the prejudice against the colour of the fur having almost disappeared. Those that are of no use for exhibition should, therefore, be fattened.

There are various colours to choose from, the most popular perhaps being the black and white, although tortoise-shells are making a big bid for premier place at the present time. The contrast in the black and white or blue and white variety is very pleasing.

The general combination of markings is very attractive, as will be seen from the following, which is typical of a black and white specimen :

The nose marking is black, and represents a butterfly with extended wings, the fork running up the centre of the face being the body of the butterfly. A circle is around each eye, with a cheek spot on either side of the face, but adjacent to the circle. Neat black ears about 4 inches in length are demanded, while the rest of the face should be spotlessly white. The " saddle " is a very important marking, and extends along the back, from the root of the ears to the tail. Chain markings should be on each side, which gradually become larger and more numerous as the loins are reached. A spot on each leg, and six black spots on the belly are necessary.

THE DUTCH

The Dutch is another very popular variety, for it is hardy and, as with the English, the does are splendid mothers. The quality of the progeny, too, can be determined in the nest.

The colour of the Dutch, whichever is bred for, is very important. Blacks must be blacks, and not rusty brown, and similarly with blues and greys, no foreign tints must be discernible. The eyes must be bright and perfectly free from " specks." This defect has caused a lot of trouble to breeders, and once "specked " or " wall " eyes are introduced into a strain it is a long and tedious work to eradicate the fault. Specimens possessing either of these defects are handicapped out of competition by up-to-date judges.

The predominating colour on the cheeks should be even and not come too near the smellers. The " blaze " is the marking which passes up the face, and tapers away as it nears the ears. The saddle is the division of the colours, and should be evenly cut right round the body. The markings on the hind feet are the stops, and should match each other, and not be more than 1½ inches in length.

These markings, along with neat ears and a sprightly little rabbit weighing about 5 lbs., make the Dutchman a favourite variety, and one that can be recommended with every confidence as likely to give good return for the time and attention expended upon it.

The Dutch also possesses some claims as a table rabbit, though the colour is against it.

THE LOP

Lop rabbits seem to be regaining some of the popularity they enjoyed ten years ago, when they were generally spoken of as the " King of the Fancy " breed. There are various colours, both self and broken, in Lops, fawns, sooty fawns, and tortoise-shells at the present time appearing to find most favour. This variety, more than any I know of, seems to be essentially suited to an indoor rabbitry, due to a great extent to the fact that heat is a necessary factor to produce the length of ear required before a specimen can command recognition in the show pen. An equable temperature of 60 degrees Fahr. must always be maintained.

No doubt the fact that it is not convenient for the majority of fanciers to possess a building suitable for a stove has been the means of many deciding against keeping Lops. The rage for length of ear has, in many cases, led to the other good points, such as size and shape, being lost in many specimens. Since, however, the best shows have put on classes for rabbits measuring a specific length, the "all property" points appear to be reasserting themselves.

There is a good demand for Lops of the right stamp, and anyone taking them up should not find any difficulty in disposing of their surplus stocks at prices to yield a profit.

The chief show point about the Lop is the ear, which should be as long as possible and of proportionate width, set low down, hang close to the cheek, and be well rounded outwards, while the texture must be fine and flexible. The head should be of fair size and nicely rounded, the eye large and full, the shoulder low with a good curve to the tail, which should be straight and carried erect.

The length of ear is ascertained by stretching out both ears and measuring from the top of one, across the forehead to the top of the other. Sometimes the length runs to as much as 26 inches.

THE SILVER

The Silver has a certain popularity as a table rabbit, but is now bred, for the most part, for show purposes, though there is always a good market for the pelts, which rank first in importance with furriers.

The recognised colours are blue, grey, fawn, and brown, and the average weight is 6 lbs., running up to 9 lbs. Of whatever shade, the chief features are a general evenness from the nose to the tail, and a very difficult matter it is to get a specimen whose feet, tail, chest, nose, and ears all match the body. Fawns and browns must possess the even ticking the same as greys to be good specimens. The ground colours are fawn and brown respectively, whilst the amount of silvering determines the shade. The brown variety is particularly hard to breed to standard requirements.

The fur must be short, springy, and thick, but not soft. A soft-coated Silver is useless from a show standpoint. The brighter and more glossy the coat and the more sparkling the silvering and ticking, the better the rabbit. Closeness of ticking and silvering is also a good point.

Whatever the colour it should be deep, rich, and even, with nose, muzzle, cheeks, feet, and tail free from dark shading or smuttiness. The ears should be short (about 3½ inches long), erect, close together, not wide at the tips, and of the same shade as the body.

The great point in breeding Silvers is to use only bucks and does in full coat, preferably a dark doe and a between-shade sire. Rabbits with rusty-coloured feet or sides are useless for breeding purposes.

THE TAN

The Tans are not a difficult breed to understand, and as they are constitutionally very hardy they can be successfully bred either in a rabbitry or hutches fixed in the open air.

Their name signifies the true colour of the breed. Nothing but black or blue and tan must appear on any part of the body. The black and the tan must reach to the skin, no foreign colour whatever being allowed in the under coat. The black colour is density itself, and most pleasing to the eye. It must be absolutely free from brindling except on the sides, and the sides of the rump, where the more brindling the better the specimen. The tan should be of a deep golden colour, absolutely free from black (or blue).

The ideal weight varies from 3 lbs. to 4½ lbs.

Taking the black variety first, the head and cheeks must be black, which colour must reach right to the nose, excepting that each eye is encircled with a ring of tan.

The shoulders (except just at the back of the ears), the back, saddle, and hindquarters, must all be black. The sides of the latter should be thickly intermixed with tan, which extends along each side of the body.

On the nostrils, chest, flanks, jowl, and belly should be found a rich mahogany tan of very bright tint. Just behind the ears is also tanned, and this colour from being broad at this point narrows over the rabbit's back and forms in shape a triangle. In a good specimen the tan forms a continuous line from this triangle near its ears and makes a collar round the rabbit's neck.

The ears are short, fine, and carried erect. The outside is jet black whilst the inside is tanned, and the more of this colour hereabouts the better.

The hind legs should all be tan colour, and the front legs black in front and tanned at the back, whilst the toes also are tan.

The same standard is applicable to Blue and Tans, substituting blue for black in every instance. The tan colouring in blues is not so rich as that to be found in blacks.

Breeders should not be in too great a hurry to dispose of young specimens they have bred. Often they apparently

I

lack some of the points mentioned, and until they are ten weeks old only experienced hands can judge with any certainty what their value is likely to be.

The tans are of a shy disposition, and will sometimes, in show pens, try to bury themselves in the hay provided for bedding. This characteristic, however, is not nearly so pronounced now as it was a few years ago, due, no doubt, to generation after generation accepting confinement as their natural mode of living.

THE HIMALAYAN

Although the Himalayan rabbit is an old variety, it still retains its popularity with many breeders, and a large number are reared every season both for exhibition and to be sold as pets. On the Continent, particularly in certain parts of France, a good trade is done with the fur obtained from this rabbit, as it has the reputation of being readily sold as imitation ermine.

The Himalayan is not a large rabbit, and cannot by any means be classed with either the Belgian Hare or Flemish Giant for utility purposes alone, but where only a few rabbits are kept, and the food supply therefrom is of secondary importance, the breed will appeal to many on account of its neat and compact appearance and pretty markings.

I recently saw a Himalayan rabbitry where the owner was emphatic that the flesh of the Himalayan was superior to any other variety. With this, however, I could not agree. On the other hand, some really good specimens have been bred and sold for exhibition by the same owner, and have yielded a remarkably good profit.

Where, therefore, one has no particular fancy for the self-coloured rabbits, and no prejudice against the marked varieties for eating purposes, the Himalayan can certainly be given a trial.

The rabbit has a short, white coat of fine texture, with the ears, nose, tail, and all four feet of chocolate brown. The denser the latter colour on the parts named the more valuable is the specimen. The eyes are pink, and make a contrast with the other markings.

Of a hardy nature, the rabbits will comfortably withstand any amount of dry cold but are very liable to sickness (as other varieties are) if exposed to draughts and wet. They will breed in outside hutches very readily, but when kept in such places they should be given a covering to protect them from the direct rays of the sun. The brown markings seem to become bleached if this is not attended to, and the rabbit appears blotchy and loses half its attractiveness.

Young Himalayans when born are all white and are three to four months old before their full value can be estimated. Reasonable care is, therefore, necessary in choosing the best specimens for sale, and on no account should too hasty an opinion be formed of the merits of the young ones until they have developed their markings.

There is a smart appearance and compactness about the Himalayan that will always command admirers, and to raise good specimens is not a hard job provided well-bred stock is used for the purpose.

THE ANGORA

Although the Angora is a recognised utility breed, as fully explained in Chapter VI, it is also extensively bred for exhibition, and, in addition, is one of the most popular pet breeds among children.

The show points of the Angora are: weight, about 6 to 7 lbs. ; ears, well covered and tufted at the ends ; fur on chest and neck as long as possible ; the head, short and well covered with fur ; eyes, a rich ruby red ; back, nicely rounded so that the fur flows gently and gracefully all round, giving the rabbit the appearance of a ball of wool.

It should be remembered that all the breeds indicated in Chapter II are found on exhibition at the rabbit show. The Beveren classes are nowadays, in fact, usually the heaviest.

In selecting inmates of one's rabbitry for the showbench preference should be given to those animals which have the greatest depth of colour, which are *uniform* in colour, and which are typical of the breed.

Faults of the Beveren, from a show point of view, are

hair or eyes other than blue, pendant ears, a coarse or woolly coat and an excessive dewlap. Any colour at all in the White Beveren is a great fault. In the Chinchilla the chief faults to look for are eyes which are any other colour than brown, barred feet, drooping or lopped ears, and white patches on feet and head.

In the Havana, an excessive dewlap is a fault, as also, of course, are white hairs. In the Argente de Champagne, white patches in the coat will disqualify an exhibit, as will drooping ears and eyes of any other colour than brown. A harsh, thin or ragged fur, a creamy tint in the fur colour, and a bony frame are serious faults.

Disqualifications in the Lilac include white patches on the body ; eyes which are some other colour than ruby red. Faults are white hairs about the pelt and a brown tint about the feet. The show faults in the Sitka are a rusty tint in the fur, an excessively pale undercolour, a small white tip to nose and toes. The eyes of a Sitka are bright dark brown.

INDEX

CPSIA information can be obtained
at www.ICGtesting.com
Printed in the USA
FSHW010855010521
81025FS